A-level Study Guide

Biology

Revised and updated for 2008 by

Gareth Rowlands

Revision E‹press

Gareth Rowlands is a teacher with over 30 years' experience in secondary schools in England and Wales. His post prior to retirement was as Head of Biology at Ysgol Syr Thomas Jones, Amlwch, Anglesey (1983–98). He also has many years' experience as an examiner and is currently a principal examiner of A-level Biology for the Welsh Joint Education Committee.

Acknowledgements

I would like to express my thanks to **Ann Rowlands** for her editorial assistance.

I am indebted to the following examination boards for permission to reproduce questions from their past examination papers:

The Welsh Joint Examining Board
The Associated Examining Board

The examination boards are not responsible for the suggested answers to the questions. The full responsibility for these is accepted by the author.

Series Consultants: Geoff Black and Stuart Wall

Project Manager: Hywel Evans

Pearson Education Limited

Edinburgh Gate, Harlow

Essex CM20 2JE, England

and Associated Companies throughout the world

© Pearson Education Limited 2000, 2003, 2008

British Library Cataloguing in Publication Data

A catalogue entry for this title is available from the British Library.

ISBN 978-1-40820649-2

First published 2000
Second edition 2003
Third edition 2008
Reprinted 2008
Set by Juice Creative Ltd
Printed and bound by Ashford Colour Press Ltd, Gosport, Hampshire

Contents

How to use this book

Specification map
Provides a quick and easy overview of the topic that you need to study for the specification you are studying

Exam themes
Gives a quick overview of the key themes that will come up in the exam

Exam themes

- The properties of biological molecules, recognizing and identifying their general formulae and structure and relating these to their roles

- The structure and replication of DNA

Action point
A suggested activity linked to the content

Checkpoint
Quick question to check your understanding with full answers given at the end of the chapter

Links
Cross-reference links to other relevant sections in the book

Examiner's secrets
Hints and tips for exam success

Watch out!
Flags up common mistakes and gives hints on how to avoid them

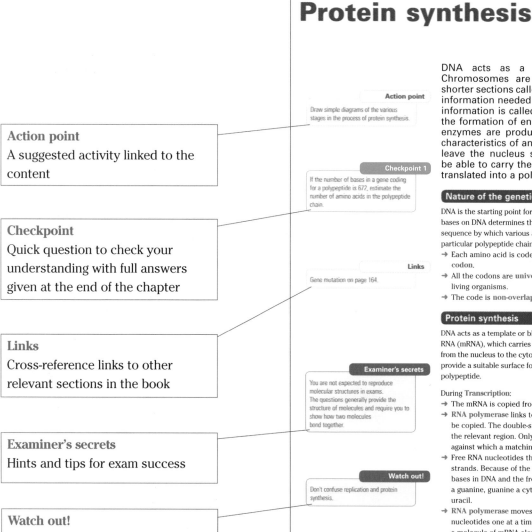

Protein synthesis

DNA acts as a store of genetic information Chromosomes are divided up into thousands o shorter sections called genes each of which carries th information needed to make a particular protein. Thi information is called the genetic code. Genes contro the formation of enzymes and by determining whic enzymes are produced the DNA can determine th characteristics of an organism. However, DNA canno leave the nucleus so a 'messenger molecule' mus be able to carry the code to the ribosome where it i translated into a polypeptide molecule.

Action point

Draw simple diagrams of the various stages in the process of protein synthesis.

Nature of the genetic code

DNA is the starting point for protein synthesis since the sequence of bases on DNA determines the primary structure of a protein, that is, the sequence by which various amino acids are joined together to form a particular polypeptide chain

→ Each amino acid is coded for by three bases – the triplet code or codon.

Checkpoint 1

If the number of bases in a gene coding for a polypeptide is 672, estimate the number of amino acids in the polypeptide chain.

→ All the codons are universal, i.e they are exactly the same for all living organisms.
→ The code is non-overlapping in that each triplet is read separately.

Links

Gene mutation on page 164.

Protein synthesis

DNA acts as a template or blueprint for the production of messenger RNA (mRNA), which carries the instructions needed for protein synthesis from the nucleus to the cytoplasm. The function of the ribosomes is to provide a suitable surface for the attachment of mRNA and the assembly o polypeptide.

During Transcription:
→ The mRNA is copied from a specific region of DNA called the cistron
→ RNA polymerase links to the DNA at the beginning of the sequence t be copied. The double-stranded DNA first unwinds and then unzips in the relevant region. Only one of the DNA strands acts as a template against which a matching mRNA can be formed.

Examiner's secrets

You are not expected to reproduce molecular structures in exams. The questions generally provide the structure of molecules and require you to show how two molecules bond together.

→ Free RNA nucleotides then align themselves opposite one of the two strands. Because of the complementary relationship between the bases in DNA and the free nucleotides, cytosine in the DNA attracts a guanine, guanine a cytosine, thymine an adenine, and adenine a uracil.

Watch out!

Don't confuse replication and protein synthesis.

→ RNA polymerase moves along the DNA forming bonds that add nucleotides one at a time to the RNA. This results in the synthesis of a molecule of mRNA alongside the unzipped portion of DNA. Behind the RNA polymerase the DNA strands rejoin to reform the double helix.

18

Topic checklist
A topic overview of the content covered and how it matches to the specification you are studying

Topic checklist

	Edexcel		AQA		OCR		WJEC	
	AS	A2	AS	A2	AS	A2	AS	A2
Molecules 1	O		O		O		O	
Molecules 2	O		O		O		O	

Revision checklists
Allow you to monitor your progress and revise the areas that you are not confident on

By the end of this chapter you should be able to:

1	Distinguish between passive and active transport.	Confident	Not confident **Revise** page 39
2	Explain the meaning of osmosis in terms of water potential.	Confident	Not confident **Revise** page 38

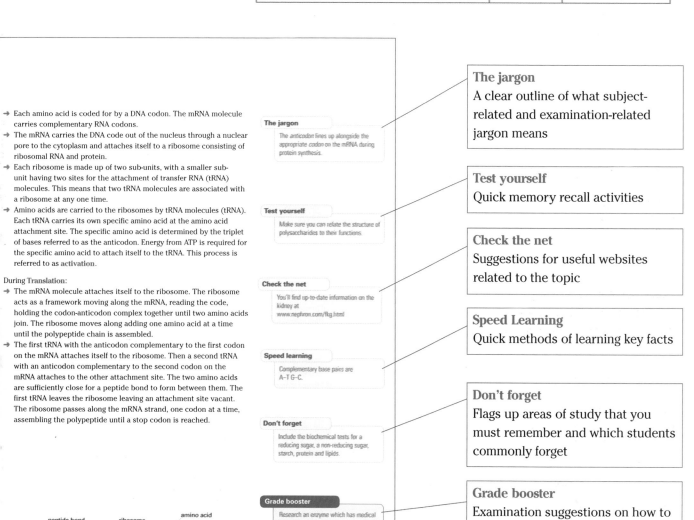

→ Each amino acid is coded for by a DNA codon. The mRNA molecule carries complementary RNA codons.

→ The mRNA carries the DNA code out of the nucleus through a nuclear pore to the cytoplasm and attaches itself to a ribosome consisting of ribosomal RNA and protein.

→ Each ribosome is made up of two sub-units, with a smaller sub-unit having two sites for the attachment of transfer RNA (tRNA) molecules. This means that two tRNA molecules are associated with a ribosome at any one time.

→ Amino acids are carried to the ribosomes by tRNA molecules (tRNA). Each tRNA carries its own specific amino acid at the amino acid attachment site. The specific amino acid is determined by the triplet of bases referred to as the anticodon. Energy from ATP is required for the specific amino acid to attach itself to the tRNA. This process is referred to as **activation**.

During **Translation:**

→ The mRNA molecule attaches itself to the ribosome. The ribosome acts as a framework moving along the mRNA, reading the code, holding the codon-anticodon complex together until two amino acids join. The ribosome moves along adding one amino acid at a time until the polypeptide chain is assembled.

→ The first tRNA with the anticodon complementary to the first codon on the mRNA attaches itself to the ribosome. Then a second tRNA with an anticodon complementary to the second codon on the mRNA attaches to the other attachment site. The two amino acids are sufficiently close for a peptide bond to form between them. The first tRNA leaves the ribosome leaving an attachment site vacant. The ribosome passes along the mRNA strand, one codon at a time, assembling the polypeptide until a stop codon is reached.

The jargon
The anticodon lines up alongside the appropriate codon on the mRNA during protein synthesis.

Test yourself
Make sure you can relate the structure of polysaccharides to their functions.

Check the net
You'll find up-to-date information on the kidney at www.nephron.com/fkg.html

Speed learning
Complementary base pairs are A–T G–C.

Don't forget
Include the biochemical tests for a reducing sugar, a non-reducing sugar, starch, protein and lipids.

Grade booster
Research an enzyme which has medical importance.

Exam practice   answers: page 32

1 Describe the functions of the various forms of nucleic acids in protein synthesis.

19

The jargon
A clear outline of what subject-related and examination-related jargon means

Test yourself
Quick memory recall activities

Check the net
Suggestions for useful websites related to the topic

Speed Learning
Quick methods of learning key facts

Don't forget
Flags up areas of study that you must remember and which students commonly forget

Grade booster
Examination suggestions on how to get the top grade

Exam practice
Exam-style questions to check your understanding of the topic content with full answers given at the end of the chapter

Specification map

Molecules and cells	Molecules 1
	Molecules 2
	Nucleic acids
	Replication and protein synthesis
	Enzymes 1
	Enzymes 2
	Cell structure and cell membrane
	Cell organelles
	Chromosome structure and mitosis
	Gene technology
Exchange and transport	Movement of molecules
	Human digestion
	Gas exchange in plants and simple organisms
	Gas exchange in mammals
	Transport in mammals: blood
	Transport in mammals: vessels
	Transport in mammals: heart
	Transport in plants
	Transport of water through the plant
	Translocation in plants
Energy and the environment	Autotrophic and heterotrophic nutrition
	Energy flow in the ecosystem 1
	Energy flow in the ecosystem 2
	Food chains and food webs
	Recycling of nutrients 1
	Recycling of nutrients 2
	Population growth and control
	Community and succession
	Biodiversity and agriculture
Metabolism, respiration and photosynthesis	The role of ATP in metabolism
	Aerobic respiration
	Anaerobic respiration and energy budgets
	Photosynthesis
	Biochemistry of the light-dependent stage
	Biochemistry of the light-independent stage
Regulation and control	Homeostasis
	The kidney
	Functioning of the kidney
	Plant growth substances
	Effect of light on plant growth
	Nerves
	Synapses
	The central nervous system
	Sensory receptors
	Muscles
Microbiology	Types of microbes
	Growth and culture techniques
	Fermentation
	Pathogens and human disease
	Protection against disease
	Applications and contemporary issues 1
	Applications and contemporary issues 2
	Applications and contemporary issues 3
Reproduction and genetics	Flower structure and gamete development
	Pollination and fertilization
	Seed development and germination
	Human reproduction
	Hormonal control of reproduction
	Fertilization and development
	Meiosis
	Variation and evolution
	Mendel's laws
	Deviations from Mendel's laws and sex linkage
	Mutations
	Human Genome Project
	Genetic counselling and gene therapy
	Genetic fingerprinting
	Applications of reproduction and genetics
	Tissue culture
	Genetically modified organisms
Classification, biodiversity and evolution	Principles of taxonomy
	The five-kingdom classification
	Population genetics
	Evolution and selection
	Isolation and speciation
	Biodiversity, extinction and conservation

Edexcel		AQA		OCR		WJEC	
AS	A2	AS	A2	AS	A2	AS	A2
○		○		○		○	
○		○		○		○	
○		○		○		○	
○	●	○	●	○	●		●
○		○		○		○	
○		○		○	●	○	
○		○		○		○	
○		○		○		○	
○		○		○		○	
○	●		●	○	●		●
○				○			
		○				○	
○		○		○		○	
○	●	○		○		○	
○		○		○		○	
○		○		○		○	
○	●	○		○		○	
○		○		○		○	
○		○		○		○	
				○		○	
	●		●		●	○	
	●		●		●		●
	●		●		●		●
	●		●		●	○	
	●		●	○			●
	●		●		●		●
	●		●		●		●
○	●	○	●	○	●		●
	●		●		●		●
	●		●		●		●
	●		●		●		●
	●		●		●		●
	●		●		●		●
	●		●		●		●
	●		●		●		●
	●		●		●		●
	●		●		●		●
	●		●		●		●
	●		●		●		●
	●		●		●		●
	●		●		●		
	●		●		●		●
	●	○		○			●
	●	○		○			●
	●	○			●		●
	●	○		○			
	●	○		○			
	●	○		○			●
		○		○			●
		○		○			●
○							●
○							●
							●
○		○			●		●
○		○					●
○		○			●		●
○	●	○	●				●
○	●	○	●	○	●		●
			●		●		●
			●		●		●
○	●		●	○	●		●
○	●		●		●		●
○	●		●		●		●
			●		●		●
○	●		●	○	●		●
○			●		●		●
○			●		●		●
○		○		○		○	
○		○		○		○	
			●		●		●
○		○		○			●
○			●	○			●
○	●	○		○		○	

7

Molecules and cells

All living organisms are made up of cells, each of which is surrounded by a membrane. These membranes are also found internally where they separate functional units, called organelles, within the cell. One important organelle is the nucleus, which contains the hereditary factors, or genes, of an organism. The manipulation of these genes is called genetic engineering. The structure and properties of biologically significant molecules such as water, carbohydrates, lipids, proteins and nucleic acids are important in understanding metabolic processes such as respiration and protein synthesis, which are controlled by biological catalysts called enzymes.

Exam themes

- The properties of biological molecules, recognizing and identifying their general formulae and structure and relating these to their roles
- The structure and replication of DNA
- The role of DNA in protein synthesis
- A knowledge of the properties of enzymes and the conditions affecting the way in which they function
- The structure and function of organelles
- The fluid mosaic model of membrane structure
- Chromosome behaviour in mitosis
- The principles of gene manipulation

Topic checklist

	Edexcel		AQA		OCR		WJEC	
	AS	A2	AS	A2	AS	A2	AS	A2
Molecules 1	O		O		O		O	
Molecules 2	O		O		O		O	
Nucleic acids	O		O		O		O	
Replication and protein synthesis	O	●	O	●	O	●		●
Enzymes 1	O		O		O		O	
Enzymes 2	O		O		O	●	O	
Cell structure and cell membrane	O		O		O		O	
Cell organelles	O		O		O		O	
Chromosome structure and mitosis	O		O		O		O	
Gene technology	O	●		●	O	●		●

Molecules 1

Even though organisms are incredibly complicated they are all made up of the same type of molecules. To understand how a molecule carries out a particular function in an organism you will need to look at the properties of the molecule. The molecule gets its properties from its structure so you must look at the structural level to understand the processes of life.

Examiner's secrets

You are not expected to reproduce molecular structures in exams. The questions generally provide the structure of molecules and require you to show how two molecules bond together.

The jargon

Polar molecules are those with unevenly distributed charges.

Hydrogen bonds are individually weak but collectively strong.

A high *specific heat* is where a large amount of heat energy is needed to cause a small rise in temperature.

A high *latent heat* is where a great deal of heat energy is needed to change from liquid state to vapour.

Water

Water is vital to all living organisms, as a constituent of cells, a reactant in metabolic reactions, a solvent and, for aquatic organisms, a habitat.

Properties of water molecules

→ Polar, with negative charge on hydroxyl ions and positive charge on hydrogen ions. The attraction of oppositely charged poles of water molecules causes them to group together. Attractive forces form hydrogen bonds.

→ High specific heat, which is important:
  → inside cells, where metabolic reactions are enzyme controlled.
  → externally to provide a constant environment for aquatic organisms.

→ High latent heat, which is significant in temperature control where heat is used for vaporization of water when sweating.

→ Maximum density at 4 °C.

→ High surface tension.

Examiner's secrets

Numbering carbon atoms from 1 to 6 in the glucose molecule is useful.

Action point

Draw diagrams of the two forms of glucose.

Checkpoint 1

How many carbons are there in a pentose sugar?

Checkpoint 2

How do α and β forms of glucose differ?

Carbohydrates

These are organic molecules containing carbon, hydrogen and oxygen. Carbohydrates are of three types: monosaccharides, disaccharides and polysaccharides. They have two main functions:

→ an easily accessible source of energy in both plants and animals

→ cellulose has a structural role in plant cell walls

Monosaccharides

These have the general formula $(CH_2O)_n$ where n is any number from 3 to 9.

$n = 3$: trioses, important as intermediates in metabolism

$n = 5$: pentoses, used in the formation of nucleic acids

$n = 6$: hexoses, e.g. glucose, the main source of energy

All sugars share the formula $C_6H_{12}O_6$ but they differ in their molecular structure, i.e. they are isomers. Glucose exists as two isomers, the α form and the β form. These different forms result in considerable biological differences when they form polymers such as starch and cellulose.

Disaccharides

Consist of two monosaccharide units linked together by the formation of a glycosidic bond *with the elimination of water*. This is a condensation reaction.

Polysaccharides

These are formed from very large numbers of monosaccharide units linked together. The two main polysaccharides to consider are:

→ starch (in plants) and glycogen (in animals)
- → they are built up of α glucose molecules
- → they act as energy stores
- → their properties include being insoluble and compact (because chains may be folded), do not alter osmotic concentrations, and do not easily diffuse out of the cell, making them ideal for storage

→ cellulose
- → consists of long chains of β glucose molecules cross-linked to each other by hydrogen bonds and these chains are grouped together into microfibrils
- → has a structural role as it gives a high tensile strength to plant cell walls

Table showing classification of selected carbohydrates.

Monosaccharides	Disaccharides	Polysaccharides
glucose	maltose	starch
fructose	sucrose	glycogen
galactose	lactose	cellulose

Action point

Draw a diagram showing the formation of a disaccharide from two monosaccharides.

Checkpoint 3

When two monosaccharides combine a molecule of water is released. Is this a hydrolysis or condensation reaction?

Action point

Make notes and draw diagrams of the structures of the different types of carbohydrates.

Test yourself

Make sure you can relate the structure of polysaccharides to their functions.

Checkpoint 4

Is cellulose made up of polymers of α or β glucose?

Action point

List the disaccharides and polysaccharides in the table and briefly describe where they are found in plants and/or animals.

Grade booster

Explain how the molecular structure of cellulose enables it to carry out its structural role in plants.

Examiner's secrets

You are rarely expected to memorize structural diagrams of polysaccharides.

Exam practice answers: page 32

1 (a) Cellulose is the most abundant naturally occurring compound on Earth. State precisely where cellulose is most likely to be found.

   (b) The diagram below shows part of a molecule of cellulose.

   (i)   Name molecule Z.

   (ii) Ring the term which correctly describes molecule Z.

         hexose   pentose   tetrose   triose

   (iii) Describe how units in the diagram are arranged in a complete molecule of cellulose.

   (iv) Name one property this molecular structure gives to cellulose.

Molecules 2

You need to study the structure of fats and proteins, both of which have important roles. Fats are particularly important in the structure of membranes. Proteins form the structural basis of all living cells.

Fats (lipids)

Lipids are **triglycerides**, composed of carbon, hydrogen and oxygen, but their oxygen content is very low. They are formed by condensation reactions between glycerol and molecules of fatty acids. They are non-polar and so are insoluble in water. Different triglyceride fats are formed from different fatty acids.

Their functions include:

→ acting as an energy store
→ acting as a heat insulator
→ protecting delicate internal organs

Phospholipids

Phospholipids are lipids in which:

→ one of the fatty acid groups is replaced by a phosphate group
→ the phosphate group is hydrophilic (polar)
→ the rest of the molecule is hydrophobic (non-polar)

They are important in the formation and functioning of membranes in cells.

Proteins

→ Proteins are built up from linear sequences of **amino acids**. About 20 different amino acids are used to make up proteins.

→ They differ from carbohydrates and lipids in that in addition to carbon, hydrogen and oxygen they always contain nitrogen. Many proteins also contain sulphur and sometimes phosphorus.

→ They possess an **amino** group (NH_2) at one end of the molecule, and a **carboxyl** group (COOH) at the other end.

Properties of proteins

Proteins are:

→ crystalline and colourless
→ amphoteric and so can act as buffers

Structure of proteins

Proteins are built up from a linear sequence of amino acids. The amino group of one amino acid reacts with the carboxyl group of another with the elimination of water (condensation reaction). The bond that is formed is called a **peptide** bond and the resulting compound is a **dipeptide**. A number of amino acids joined in this way is called a **polypeptide**.

Check the net

You'll find up-to-date information about proteins at www.indiana.edu/~cheminfo/09-16.html

Action point

Draw a diagram to show the formation of a triglyceride from one molecule of glycerol and three molecules of fatty acid.

Checkpoint 1

A lipid is insoluble in water. How does the addition of a phosphate group affect this property?

The jargon

Hydrophilic means to attract water. *Hydrophobic* means to repel water.

Links

Look at the structure of the cell membrane on pages 24 and 25.

Checkpoint 2

Why are the molecules of an amino acid described as polar?

Test yourself

Write down the generalized structural formula of an amino acid.

The jargon

An *amphoteric* substance can act as both an acid and a base.
Buffers resist changes in pH.

Action point

Draw a diagram to show how two amino acids join together to form a dipeptide.

Four levels of protein structure exist.

1 The **primary** structure of a protein is the sequence of amino acids in its polypeptide chain. The proteins differ from each other in the variety, numbers and orders of their constituent amino acids.

2 The **secondary** structure is the shape that the polypeptide chain forms as a result of hydrogen bonding. This is most often a spiral known as the α helix. An alternative is a pleated sheet occurring as a flat zig-zag chain.

3 The **tertiary** structure is formed by the bending and twisting of the polypeptide helix into a compact three-dimensional (3D) structure. The shape is maintained by disulphide, ionic and hydrogen bonds.

4 The **quaternary** structure arises from a combination of a number of different polypeptide chains. These are associated with non-protein groups and form large, complex molecules, e.g. a haemoglobin molecule is formed from four separate polypeptide chains, each associated with a complex iron-containing prosthetic group, called haem.

Action point

Draw sketches of the different levels of protein structure.

Checkpoint 3

How is the secondary structure of a protein maintained?

Checkpoint 4

How are ionic and hydrogen bonds affected by heat?

Grade booster

Be prepared to answer questions which link the structure of carbohydrates, fats and proteins.

Classification of proteins

Proteins can be classified according to their structure:

→ **fibrous** proteins (perform structural functions) are
  → insoluble in water
  → strong and tough
  → made up of long polypeptide chains or sheets often with numerous cross-linkages

→ **globular** proteins (function as enzymes, antibodies, plasma proteins and hormones) are
  → folded as globular complex 3D molecules
  → soluble in water

Checkpoint 5

What type of proteins are enzymes?

Action point

Make a list of the uses of proteins. You'll be surprised how many uses there are!

Grade booster

Relate the secondary and tertiary levels of protein structure to their function.

Don't forget

Include the biochemical tests for a reducing sugar, a non-reducing sugar, starch, protein and lipids.

Exam practice answers: page 32

1 The diagram below shows the structural formula of a typical amino acid.
  (a) (i) Name the parts of the molecule labelled A and B.

$$H_2N + C + COOH$$

with A above H_2N, B above $COOH$, R above C, and H below C.

  (ii) State what R would represent in the simplest amino acid.
  (b) (i) Draw a diagram to show two molecules of amino acids joined by means of a peptide bond.
    (ii) Label the peptide bond on your diagram.
    (iii) Name the type of reaction involved when two amino acids combine in this way.

Nucleic acids

Check the net

You'll find multiple choice questions on nucleic acids at www.biology.arizona.edu

Nucleic acids are the molecules that contain genetic information and are found in all living cells and viruses. Two types of nucleic acid are found in cells: deoxyribonucleic acid (DNA) and ribonucleic acid (RNA). The structure of DNA was worked out by Watson and Crick in the 1950s.

Structure of nucleic acids

Nucleic acids are built up of units called **nucleotides** (see diagram below).

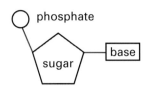

Individual nucleotides are made up of three parts which combine by condensation reactions. These are:

→ phosphoric acid (phosphate H_3PO_4)
→ pentose sugar, of which there are two types
  → in RNA the sugar is ribose
  → in DNA the sugar is deoxyribose
→ organic base, of which there are five, divided into two groups
  → the **pyrimidine** bases are thymine, cytosine and uracil
  → the **purine** bases are adenine and guanine

The jargon

Nitrogenous bases derived from *pyrimidines* are single ring structures whereas those derived from *purines* are double ring structures.

The jargon

A *polymer* is a large number of repeating identical units.

DNA

The structure is as follows:

→ it is a double-stranded polymer of nucleotides (polynucleotide)
→ each polynucleotide may contain many million nucleotide units
→ it is in the form of a **double helix**, the shape of which is maintained by hydrogen bonding
→ the pentose sugar is always deoxyribose
→ it contains four organic bases, adenine, guanine, cytosine and thymine
→ each strand is linked to the other by pairs of organic bases
→ cytosine always pairs with guanine, adenine always pairs with thymine, and the bases are joined by hydrogen bonds

Speed learning

Complementary base pairs are A–T G–C.

Checkpoint 1

Write out the complementary sequence of bases to AGCCTACGT.

Grade booster

Interpret data from experimental work investigating the role of nucleic acids.

This diagram shows how nucleotides are arranged relative to each other in part of a DNA chain.

complementary base pairs

sugar phosphate backbone

hydrogen bonds

RNA

→ RNA is a single-stranded polymer of nucleotide.
→ It contains the pentose sugar, ribose.
→ It contains the organic bases adenine, guanine, cytosine and uracil.

There are three types of RNA.

→ *Ribosomal RNA (rRNA)* is found in the cytoplasm and is a large complex molecule made up of both double and single helices.
→ *Transfer RNA (tRNA)* is a small single-stranded molecule. It forms a clover-leaf shape, with one end of the chain ending in a cytosine–cytosine–adenine sequence where the amino acid it carries attaches itself. At the opposite end of the chain is a sequence of three bases called the anticodon (see diagram below).
→ *Messenger RNA (mRNA)* is a long single-stranded molecule, formed into a helix. It is manufactured in the nucleus and passes into the cytoplasm where it associates with the ribosomes.

Action point

Make your own notes and diagrams on the three different types of RNA.

Links

Protein synthesis is found on pages 18 and 19.

Watch out!

In RNA uracil replaces thymine, so A pairs with U.

Examiner's secrets

You can expect to have a question on some aspect of DNA structure.

The jargon

The *anticodon* lines up alongside the appropriate *codon* on the mRNA during protein synthesis.

Checkpoint 2

If DNA is likened to a 'ladder' with alternating phosphate and deoxyribose molecules making up the 'uprights', what are the 'rungs' made up of?

Checkpoint 3

State two ways in which the structure of RNA differs from that of DNA.

Checkpoint 4

Which of RNA and DNA remains constant for all cells (except gametes) of a species?

Test yourself

Draw up a table of differences between RNA and DNA.

Exam practice   answers: page 32

1 The diagram below represents a small portion of DNA.

(a) Name:

    (i)   the bond labelled X

    (ii) one base which could be represented by Z

    (iii) two elements other than carbon, hydrogen and oxygen which are present in DNA

(b) Describe three ways in which the molecular structure of DNA differs from mRNA.

Replication

Chromosomes must make copies of themselves so that when cells divide, each daughter cell must receive an exact copy of the genetic information. This copying of DNA is called replication and takes place in a cell during interphase.

Links

Look at the section on mitosis on pages 28 and 29.

The jargon

Complementary bases are the bases that pair with each other, e.g. A–T, G–C.

The jargon

A *template* is a blueprint from which a copy is made.

Checkpoint 1

Why are the bacteria washed thoroughly before being transferred from the medium containing ^{15}N?

Checkpoint 2

Name the two scientists who produced evidence to support the semi-conservative hypothesis.

Checkpoint 3

If generation 1 contained 50% ^{15}N and 50% ^{14}N, predict the proportions in the second generation.

Checkpoint 4

At the end of the first generation there will be 2 molecules of DNA. How many molecules of DNA will there be at the end of four generations?

Grade booster

Provide evidence for a hypothesis.

Replication

Replication occurs as follows:

→ The hydrogen bonds holding the base pairs together break and the two halves of the molecule separate.

→ **DNA unwinds** and as the strands separate **DNA polymerase** catalyses the addition of free nucleotides to the exposed bases.

→ Each chain acts as a **template** so that free nucleotides can be joined to their **complementary bases** by the enzyme, DNA polymerase.

→ The result is two DNA molecules, each made up of one newly synthesised chain and one chain that has been conserved from the original molecule.

→ This is called the **semi-conservative hypothesis**. Meselson and Stahl carried out experiments to confirm the earlier theory proposed by Watson and Crick (1954).

The experiments involved the use of an ultra-centrifuge. This rotates centrifuge tubes containing liquid suspensions at very high speeds which results in the denser particles separating out at a lower point in the tube than the lighter particles.

→ The scientists cultured the bacterium, *Escherichia coli*, for several generations on a medium containing amino acids made with the heavy isotope of nitrogen ^{15}N. The bacteria incorporated the ^{15}N into their nucleotides and then into their DNA so that all the DNA contained ^{15}N. They extracted the bacterial DNA and centrifuged it. The DNA settled at a low point in the tube.

→ The ^{15}N bacteria were washed, then transferred to a medium containing the normal, lighter form of nitrogen, ^{14}N, and were allowed to divide once more.

→ When extracts of DNA from this first generation culture were centrifuged it was shown to have a mid point density, since half the strand was made up of the original strand of ^{15}N DNA and the other half was made up of the new strand containing ^{14}N.

→ When extracts were taken from the second generation grown in ^{14}N the DNA settled at mid points and high points in the tube. This was conclusive evidence for the semi-conservative hypothesis.

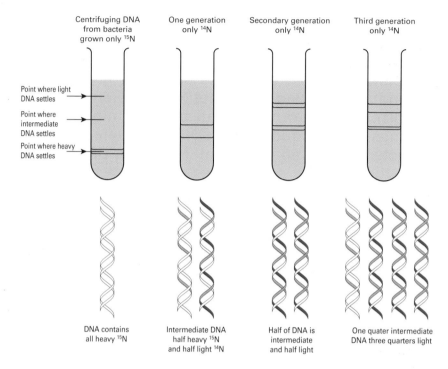

Centrifuging DNA from bacteria grown only 15N

One generation only 14N

Secondary generation only 14N

Third generation only 14N

Point where light DNA settles

Point where intermediate DNA settles

Point where heavy DNA settles

DNA contains all heavy 15N

Intermediate DNA half heavy 15N and half light 14N

Half of DNA is intermediate and half light

One quater intermediate DNA three quarters light

answers: page 32

Exam practice

1 The bacterium *E.coli* was cultured in a growth medium containing the heavy isotope of nitrogen 15N. The cells were transferred to a medium containing normal nitrogen 14N and allowed to divide twice. The DNA was extracted from these cells and the diagram shows its distribution after centrifugation.

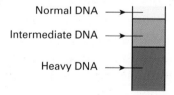

Normal DNA →

Intermediate DNA →

Heavy DNA →

(a) What would be the proportion of the two types of DNA in the second generation?

(b) Explain how you arrived at your answer.

(c) Name the enzyme mainly responsible for the synthesis of the new strand of DNA.

(d) Give an example of the practical application of this enzyme.

Protein synthesis

DNA acts as a store of genetic information. Chromosomes are divided up into thousands of shorter sections called genes each of which carries the information needed to make a particular protein. This information is called the genetic code. Genes control the formation of enzymes and by determining which enzymes are produced the DNA can determine the characteristics of an organism. However, DNA cannot leave the nucleus so a 'messenger molecule' must be able to carry the code to the ribosome where it is translated into a polypeptide molecule.

Nature of the genetic code

DNA is the starting point for protein synthesis since the sequence of bases on DNA determines the primary structure of a protein, that is, the sequence by which various amino acids are joined together to form a particular polypeptide chain

→ Each amino acid is coded for by three bases – the triplet code or **codon**.

→ All the codons are **universal**, i.e they are exactly the same for all living organisms.

→ The code is **non-overlapping** in that each triplet is read separately.

Protein synthesis

DNA acts as a template or blueprint for the production of messenger RNA (mRNA), which carries the instructions needed for protein synthesis from the nucleus to the cytoplasm. The function of the ribosomes is to provide a suitable surface for the attachment of mRNA and the assembly of polypeptide.

During **Transcription**:

→ The mRNA is copied from a specific region of DNA called the cistron.

→ **RNA polymerase** links to the DNA at the beginning of the sequence to be copied. The double-stranded DNA first unwinds and then unzips in the relevant region. Only one of the DNA strands acts as a template against which a matching mRNA can be formed.

→ Free RNA nucleotides then align themselves opposite one of the two strands. Because of the complementary relationship between the bases in DNA and the free nucleotides, cytosine in the DNA attracts a guanine, guanine a cytosine, thymine an adenine, and adenine a uracil.

→ **RNA polymerase** moves along the DNA forming bonds that add nucleotides one at a time to the RNA. This results in the synthesis of a molecule of mRNA alongside the unzipped portion of DNA. Behind the RNA polymerase the DNA strands rejoin to reform the double helix.

Links

Gene mutation on page 170.

Watch out!

Don't confuse replication and protein synthesis.

Checkpoint 1

If the number of bases in a gene coding for a polypeptide is 672, estimate the number of amino acids in the polypeptide chain.

Checkpoint 2

In which organelle does protein synthesis occur?

Checkpoint 3

Name the molecule that moves through the nuclear pore and attaches to a ribosome.

The jargon

Each tRNA differs in the composition of a triplet of bases called the *anticodon*.

Checkpoint 4

Which molecule carries an amino acid to a ribosome?

Checkpoint 5

Name the molecule which is transcribed, but not translated.

→ Each amino acid is coded for by a DNA codon. The mRNA molecule carries complementary RNA codons.

→ The mRNA carries the DNA code out of the nucleus through a nuclear pore to the cytoplasm and attaches itself to a ribosome consisting of ribosomal RNA and protein.

→ Each ribosome is made up of two sub-units, with a smaller sub-unit having two sites for the attachment of transfer RNA (tRNA) molecules. This means that two tRNA molecules are associated with a ribosome at any one time.

→ Amino acids are carried to the ribosomes by tRNA molecules (tRNA). Each tRNA carries its own specific amino acid at the amino acid attachment site. The specific amino acid is determined by the triplet of bases referred to as the anticodon. Energy from ATP is required for the specific amino acid to attach itself to the tRNA. This process is referred to as **activation**.

During **Translation**:

→ The mRNA molecule attaches itself to the ribosome. The ribosome acts as a framework moving along the mRNA, reading the code, holding the codon-anticodon complex together until two amino acids join. The ribosome moves along adding one amino acid at a time until the polypeptide chain is assembled.

→ The first tRNA with the anticodon complementary to the first codon on the mRNA attaches itself to the ribosome. Then a second tRNA with an anticodon complementary to the second codon on the mRNA attaches to the other attachment site. The two amino acids are sufficiently close for a peptide bond to form between them. The first tRNA leaves the ribosome leaving an attachment site vacant. The ribosome passes along the mRNA strand, one codon at a time, assembling the polypeptide until a stop codon is reached.

Action point

Draw simple diagrams of the various stages in the process of protein synthesis.

Checkpoint 6

Which molecule is the site of the anticodon?

Checkpoint 7

Name the molecule that is the site of the codon.

Checkpoint 8

Why is the code a triplet code?

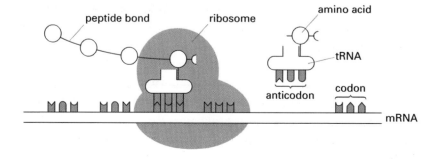

Examiner's secrets

There is a good chance that there will be a question on protein synthesis. You may be asked a question about a specific part of the process, e.g. to distinguish between transcription and translocation; to decode base sequences; or to write an essay on the complete process.

Exam practice   answers: page 32

1 Describe the functions of the various forms of nucleic acids in protein synthesis.

Enzymes 1

Metabolic reactions occur quickly and thousands of reactions are taking place at the same time in cells. Order and control is essential if reactions are not to interfere with each other. These features of metabolism are made possible by the action of enzymes. For centuries humans have used enzymes to produce bread, cheese, yoghurt, wine and beer. But now they are used on a wide commercial scale in the pharmaceutical and agrochemical industries and for analytical purposes.

Links

Digestive enzymes on page 40.
There is a question on 'immobilized enzymes' in the Resources section.

Checkpoint 1

Why is catalase called the fastest enzyme in the west?

Action point

Draw a graph to illustrate how enzymes reduce the level of activation energy in a chemical reaction.

Properties of enzymes

Enzymes have the following properties:
→ are large protein molecules
→ act as catalysts
→ lower activation energy
→ are reversible in their action
→ are very efficient, having high turnover numbers
→ can be denatured by heat
→ most are specific

How enzymes work

Enzymes react with another molecule called a **substrate**. Each enzyme has its own special shape, with an area (the **active site**) on to which the substrate molecules bind (see diagram below).

Checkpoint 2

What is meant by the term 'active site'?

Test yourself

With the aid of diagrams, explain the 'lock and key' theory of enzyme action.

enzyme + substrate

enzyme substrate complex

enzyme + products

Checkpoint 3

How does temperature affect the rate of molecular collisions?

Modern interpretations of the lock and key theory suggest that in the presence of the substrate the active site may change in order to select the substrate's shape. This is called the induced fit hypothesis.

Checkpoint 4

If all the active sites of the enzyme are occupied, what is the effect of increasing the substrate concentration?

Factors affecting enzyme action

Various factors influence the action of enzymes:

→ temperature
   → many enzymes are denatured by temperatures above 45°C
   → their optimum temperature is about 37°C
   → at low temperatures, e.g. freezing, enzymes work very slowly
→ pH – individual enzymes have a narrow optimum range
→ substrate concentration – the Michaelis constant is the concentration of substrate needed to make the reaction proceed at half its maximum rate. A low Michaelis constant means the reaction proceeds rapidly because the enzyme and substrate have a high affinity for each other.

Action point

Draw graphs to illustrate these factors.

The jargon

Affinity means chemical attraction.

The jargon

Immobilized enzymes are trapped and held in a framework of cellulose which is permeable to the substrate and products of the reaction.

Enzyme cofactors

A cofactor is a non-protein substance that is essential for some enzymes to function efficiently. Cofactors may be metallic ions, or low molecular weight organic molecules that are either temporarily or permanently attached to the enzyme molecule. They are of three types: **activators**, **coenzymes** and **prosthetic groups**.

Uses of enzymes

Enzymes are used in:

→ the food industry, e.g. pectins in plant cell walls. When fruits are mashed, pectins cause changes, including colour and flavour deterioration. When fruit ripen, pectinases are produced causing the fruit to soften. Fruit juice manufacturers add industrial pectinase between mashing and pressing to break down pectin so that a good quality, clear juice is obtained.

→ the chemical industry, e.g. proteases are used in biological detergents.

→ in medicine, e.g. in biosensors. The specificity and sensitivity of enzymes is made use of in enzyme-based devices. Biosensors can be used for the rapid and accurate detection of minute traces of biologically important molecules:
　→ in pregnancy testing kits
　→ monitoring blood sugar levels in diabetics

The enzyme is immobilized on to the surface of an electrode by being enclosed in small alginate beads. It may be used to measure quantities of urea in blood or urine samples (see diagram below). The electrode can detect changes in substrate or product, temperature changes or optical properties.

hand-held biosensor for urea

- display
- amplifier
- transducer
- gel containing immobilized urease (biological recognition layer of sensor)
- membrane permeable to urea

Immobilized enzymes are also used widely in industrial processes, such as fermentation, as they can readily be recovered for reuse. Biosensors are also used in fermenters to provide a rapid, sensitive and specific measurement of products.

Examiner's secrets

Synoptic questions often include enzymes as they are involved in many aspects of biology.

Action point

Write your own notes on other uses of enzymes.

The jargon

Biosensor describes the association of a *biomolecule*, such as an enzyme, with a *transducer*, which produces an electrical signal in response to substrate transformation.

Action point

Write brief notes on additional methods of immobilizing enzymes.

Checkpoint 5

State one advantage of immobilized enzymes other than making them easier to reuse than 'free' enzymes.

Checkpoint 6

How do enzymes differ from industrial inorganic catalysts?

Exam practice

answers: page 33

1 Explain fully why almost all biological reactions are catalyzed by enzymes and show how the structure of enzymes is adapted to this role.

Links

Fermentation is dealt with in detail on pages 132 and 133.

Enzymes 2

The actions of enzymes may be enhanced or inhibited by various substances, some found in the cell and others that are absorbed from the external environment. Certain drugs and pesticides alter metabolism by enzyme inhibition.

Watch out!

Don't confuse the different types of inhibitors.

Checkpoint 1

If a mixture is prepared containing an enzyme, a competitive inhibitor and a small quantity of substrate, what would be the probable effect on the rate of reaction if more substrate is added?

Test yourself

Draw a similar diagram to show how a non-competitive inhibitor affects enzyme action.

Checkpoint 2

In terms of attachment to the enzyme what is the difference between competitive and non-competitive inhibitors?

Checkpoint 3

What type of inhibition is affected by the concentration of the substrate?

Checkpoint 4

In terms of definition what is the main difference between reversible and non-reversible inhibition?

The jargon

Metabolites are the molecules involved in metabolic reactions.

Enzymes and inhibitors

The rate of enzyme-controlled reactions may be decreased by the presence of inhibitors. There are two main types of inhibition, **reversible** and **non-reversible**.

Reversible inhibition

The effect of this type of inhibitor is temporary. There is no permanent damage to the enzyme because the association of the inhibitor with the enzyme is a loose one. Removal of the inhibitor allows the enzyme to function normally. There are two types of reversible inhibitors.

→ *Competitive*, where the inhibitor is structurally similar to the substrate. The more substrate molecules the greater the chance of finding active sites, leaving fewer to be occupied by the inhibitor (see diagram below).

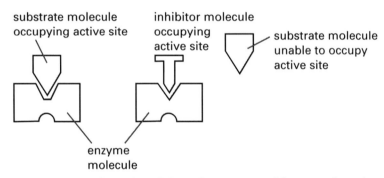

→ *Non-competitive*, where the inhibitor has no resemblance to the substrate molecule and binds to the enzyme at a site away from the active site. The rate of reaction is therefore unaffected by substrate concentration.

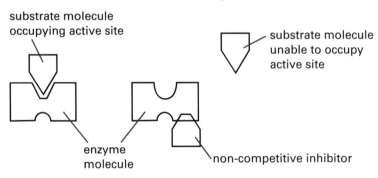

Non-reversible inhibition

These leave the enzyme permanently damaged, e.g. mercury breaks the disulphide bonds which maintain the shape of the enzyme molecule.

End-product inhibition

In some metabolic pathways the end-product of the pathway may act as an inhibitor. This end-product inhibition is an example of a negative feedback mechanism, preventing the unnecessary accumulation of a metabolite.

Allosteric effectors

These are substances present in cells that reversibly bind with the enzyme away from the active site yet cause a change in the structure of the active site. They are of two types:

→ allosteric activators, which speed up catalysis

→ allosteric inhibitors, which slow down the reaction and provide a way of controlling enzyme activity in metabolism

Grade booster

Research an enzyme which has medical importance.

Links

Enzymes in protein synthesis, respiration and photosynthesis.

Grade booster

Metabolic poisons may be enzyme inhibitors. Describe the action of one named poison.

Exam practice   answers: page 33

1   The graph below shows the relationship between substrate concentration and the initial rate of an enzyme-catalyzed reaction under different conditions.

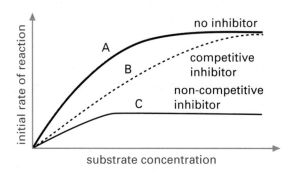

(a) Explain:

   (i)   the shape of curve A

   (ii) the difference between the shapes of curve B and curve C

(b) (i) Explain what is meant by the induced fit model of enzyme action.

   (ii) Suggest how this may provide a better explanation for the effects of a non-competitive inhibitor than the lock and key model.

2   The thermostability of enzymes is important in industrial use. One such enzyme is papain. The table below compares the thermostability of the enzyme papain in soluble and in immobilized forms.

Temperature   *Rate of reaction compared with rate at 25 °C (%)*

(°C)	Soluble ('free')	Immobilized
30	100	100
40	100	100
50	90	100
60	78	90
70	38	78
80	10	50

(a) Explain in terms of protein structure why enzymes are inactivated by high temperatures.

(b) Using the table give evidence that the thermostability of papain is increased by immobilizing it.

(c) Briefly describe one method which might have been used to immobilize the enzyme.

Examiner's secrets

Take your time and study the information provided. The graphs and tables give you clues to the answers expected.

Cell structure and cell membrane

The cell is the basic structural and functional unit of an organism. The distinction between eukaryotic and prokaryotic cells is one of the most fundamental dividing lines between living organisms. The effective boundary between the cell and its environment is the cell membrane.

Prokaryotic and eukaryotic cells

Prokaryotic cells were probably the first forms of life on Earth. This type of cell is found in bacteria and blue-green algae. Their DNA is not enclosed within a nuclear membrane and there are no membrane-bound organelles.

Eukaryotic cells are typical of the great majority of organisms including all animals and plants. They possess membrane-bound organelles such as mitochondria and chloroplasts, and a distinct membrane-bound nucleus.

Ultrastructure of the cell

The electron microscope enables the fine structure of cells and their organelles to be seen, as shown below.

Cell membrane

The cell membrane is made up almost entirely of proteins and phospholipids. The fluid mosaic model was suggested by Singer and Nicholson (see diagram). They proposed that:

→ there is a bimolecular phospholipid layer with inwardly directed hydrophobic lipid tails and outwardly directed hydrophilic phosphate heads

→ associated with the bilayer is a variety of protein molecules
  → some of the proteins occur on or in only one of the layers (extrinsic proteins)
  → some proteins extend across both layers (intrinsic proteins)

→ the phospholipid layer can move, i.e. it is fluid and in surface view the proteins are dotted throughout the layer in a mosaic arrangement

---

Checkpoint 1

Construct a table of differences between eukaryote and prokaryote cells.

Checkpoint 2

What term is applied to the outer layer of an animal cell?

Checkpoint 3

What additional outer layer is found in plant cells only?

Test yourself

Explain in your own words why Singer and Nicholson's theory is aptly named.

Links

Cell organelles are described on pages 26 and 27, and phospholipids are described on page 12.

The jargon

Bimolecular in this context means phospholipid molecules stretched out to form two layers.

exterior of cell

branching carbohydrate attached to protein to form glycoprotein

fibrous protein

phospholipid layer

extrinsic protein

intrinsic protein

cholesterol

channel-forming proteins

interior of cell

The main **functions** of the cell membrane include:

→ structural support
→ secreting chemicals
→ cell – cell recognition
→ taking up nutrients and other requirements

The membrane as a barrier

The cell surface membrane is selectively permeable to water and some solutes. Lipid-soluble substances can move through the cell membrane more easily than water-soluble substances.

→ Small, uncharged molecules, such as oxygen and carbon dioxide, freely pass through the membrane as they are soluble in the lipid part.
→ Lipid soluble molecules such as glycerol can pass through the membrane.
→ The hydrophobic core of the membrane impedes the transport of ions and polar molecules.
→ Charged particles (ions) and relatively large molecules such as glucose cannot diffuse across the non-polar centre of the phospholipid bilayer because they are relatively insoluble in lipid. Intrinsic proteins assist such particles to pass in or out of the cell by a passive process called **facilitated diffusion**. These proteins are of two types: channel proteins and carrier proteins.

Checkpoint 4

What is the difference between extrinsic and intrinsic proteins found in cell membranes?

Checkpoint 5

Why is the membrane described by the fluid mosaic model referred to as a 'dynamic structure'?

Checkpoint 6

What type of molecules freely cross the cell membrane?

Checkpoint 7

How does water cross the membrane barrier?

Links

Facilitated diffusion and active transport are dealt with in detail on pages 38 and 39 respectively.

Action point

Make a table to show the differences between plant and animal cells.

Grade booster

Research the role of cholesterol in the membrane of animal cells. Is there a link between blood cholesterol levels and heart disease?

Exam practice   answers: page 33

1 With reference to the cell membrane explain fully what is meant by each of these two terms:

(a) bilayer structure

(b) fluid mosaic

Cell organelles

Eukaryotic cells have organelles bounded by membranes. All the cell organelles are contained within a cytoplasmic matrix or cytosol. This is an aqueous solution of many vital cellular molecules. The cytosol holds and connects the various cell organelles.

Nucleus

This is the most prominent feature in the cell. Its function is to control the cell's activities and to retain the chromosomes. It is bounded by a double membrane, the **nuclear envelope**. The cytoplasm-like material within the nucleus is called the nucleoplasm. Within the nucleus are one or two small spherical bodies, each called a nucleolus.

Rough and smooth endoplasmic reticulum

The cytoplasm is a highly organized material consisting of a soluble ground substance called the **cytosol**. This contains an elaborate system of parallel, flattened cavities lined with a thin membrane. This system is known as the **endoplasmic reticulum** (ER). The cavities are interconnected and the lining membranes are continuous with the nuclear envelope. There are two types of ER:

→ *rough ER*, where the membranes are lined with ribosomes, the sites of protein synthesis
→ *smooth ER*, where the membranes lack ribosomes, being concerned with the synthesis and transport of lipids

Golgi apparatus

This is similar in structure to ER but is more compact. The Golgi body is formed by rough ER being pinched off at the ends to form small vesicles. A number of these vesicles then join up and fuse to form the Golgi body. Proteins are transported in the vesicles and are modified and packaged in the Golgi body. For example proteins may be combined with carbohydrates to make glycoproteins. At the other end of the Golgi body vesicles can be pinched off and the products secreted by exocytosis when the vesicle moves to and fuses with the cell membrane.

Chloroplast

The chloroplast is found in plant cells only. It is bounded by a double membrane. Within the chloroplast are two regions:

→ the **stroma**, which is a colourless gelatinous matrix in which the grana are embedded
→ the **grana**, each made up of a number of closed flattened sacs called **thylakoids**, within which are found the photosynthetic pigments such as chlorophyll

Mitochondrion

The mitochondrion is bounded by two thin membranes separated by a narrow fluid-filled space. The inner membrane is folded inwards to form extensions called **cristae**. The interior contains an organic matrix containing numerous

Action point

Draw up a table relating structure to function for the various organelles.

Checkpoint 1

In the context of protein synthesis what is the most likely role of the nuclear pores?

Checkpoint 2

Which organelle has a double membrane and is found only in plant cells?

Action point

Describe the processes of endocytosis and exocytosis.

Links

You will need to study chloroplasts and mitochondria in detail in order to understand the biochemistry of photosynthesis and respiration described in the section on page 92.

Check the net

You'll find up-to-date information on the mitochondrion at cellbio.utmb.edu/cellbio/mitoch1.htm

Checkpoint 3

Which organelle is concerned with respiration?

chemical compounds. The chemical reactions of aerobic respiration take place in the mitochondrion. Some of the reactions take place in the matrix, while others occur on the inner membrane and in the cytosol. The cristae increase the surface area on which the respiratory processes take place.

Lysosomes

The lysosomes contain and isolate digestive enzymes from the remainder of the cell. Digestion is carried out in a membrane-lined vacuole into which several lysosomes may discharge their contents. One of their functions is to destroy worn out organelles in the cell.

Ribosomes

These are made up of ribosomal RNA and protein. They are important in protein synthesis.

Centrioles

Centrioles are found in all animal cells and most protoctists but are absent from the cells of higher plants. Centrioles arise in a distinct region of the cytoplasm known as the **centrosome** and consist of two hollow cylinders. At cell division they migrate to opposite poles of the cell where they synthesize the microtubules of the spindle.

Microtubules

These are slender, unbranched tubes made of protein. They have several functions, e.g. they can assemble to form the spindle in cell division.

Vacuoles

Vacuoles are fluid-filled cavities bounded by a single membrane. The mature plant cell contains a large, permanent, central vacuole filled with chemicals such as glucose and ions in water. This is called the **cell sap**. The membrane around the vacuole is called the **tonoplast**. The vacuoles of animal cells are much smaller and less permanent.

Cellulose cell wall

The cell wall consists of cellulose microfibrils embedded in a polysaccharide matrix. The main functions of a cell wall are:
→ to provide strength and support
→ to permit the movement of water from cell to cell

Checkpoint 4

Suggest the difference in destination of proteins produced by 'free' ribosomes and those produced by rough ER.

Links

Details of cell division (mitosis) are on pages 28 and 29.

Links

The vacuole plays an important role in maintaining the internal hydrostatic pressure of the cell as described on page 39.

Links

Review the structure of cellulose on page 11.

Grade booster

Find out how the various organelles are linked, e.g. nucleus and ribosomes; rough ER and Golgi body.

Exam practice   answers: page 34

1 The diagram below shows part of an animal cell and is based on a series of electron micrographs.
   (a) Name the structures labelled A–D.
   (b) Label with a letter E a structure where the protein contents of D are synthesized.

Chromosome structure and mitosis

You've already learnt that inside the nucleus are chromosomes. Chromosomes contain DNA, which contains hereditary information that is transmitted from cell to cell when cells divide. In any one species the number of chromosomes per body cell nucleus is constant. Mitosis is the division of the nucleus to produce two daughter nuclei containing identical sets of chromosomes.

Check the net

You'll find up-to-date information on mitosis at
buglady.clc.uc.edu/biology/bio104/mitosis.htm
www.biology.arizona.edu

Checkpoint 1

Name the point at which pairs of chromatids are held together.

Examiner's secrets

Be prepared to compare meiosis and mitosis, and don't confuse the two. (Meiosis is described on pages 162 and 163.)

Speed learning

Memorize the correct order of the stages of mitosis by making up a mnemonic, e.g. Important People Make Awful Tea.

Action point

Construct a table summarizing the changes that occur at each of the four stages of mitosis.

Checkpoint 2

Name the stage at which the DNA content doubles.

Chromosome structure

Chromosomes are made up of DNA, protein and a small amount of RNA. DNA occurs as a single strand in the form of a double helix running the length of the chromosome. Each DNA molecule is made up of many small sections called **genes**. It is only at the onset of cell division that chromosomes become visible. Shortly before cell division begins each DNA molecule makes a copy of itself. The single thread of DNA becomes two identical threads. These are called **chromatids** and they lie parallel along most of their length but are joined only in a specialised region called the **centromere**. The number of chromosomes in the cells of different species varies. Humans always have 46 chromosomes, a fruit fly has 8 chromosomes, a potato 48 chromosomes! Chromosomes are found in matching pairs, called homologous pairs. So humans have 23 pairs of homologous chromosomes. The total number of chromosomes is called the diploid number. Sex cells, or **gametes**, have half the diploid number, this is called haploid.

Mitosis

Dividing cells undergo a regular pattern of events, known as the cell cycle. This is a continuous process but for convenience of description it is subdivided into four stages plus a 'resting' stage, known as interphase, between one complete division.

Interphase

This is the longest part of the cycle during which a newly formed cell increases in size and produces organelles lost during the previous division. Just before the next cell division the chromosomes replicate so that each then consists of two chromatids joined together by the centromere.

Prophase

This is the phase of mitotic division during which the chromatids shorten and become thicker by the coiling and condensation of the DNA protein coat. In the cells of animals and lower plants, the centrioles move to the poles of the cells and microtubules begin to radiate from them forming asters. At the end of prophase the nuclear membrane disintegrates and the spindle is formed. During this phase the nucleoli disappear.

Metaphase

The pairs of chromatids are arranged at the centre or equator of the spindle and become attached to certain spindle fibres at the centromere. Contraction of these fibres draws the individual chromatids slightly apart.

Anaphase

This stage is very rapid. The centromere splits and the spindle fibres pull the now separated chromatids to the poles of the cell, where they become the chromosomes of the two daughter cells.

Telophase

Mitosis ends with telophase. The chromosomes, having reached the poles of the cells, uncoil and lengthen. The spindle breaks down, the centrioles replicate, the nucleoli reappear and the nuclear membrane re-forms. In animal cells **cytokinesis** occurs by the constriction of the centre of the parent cell from the outside inwards. In plant cells, a cell plate forms across the equator of the parent cell from the centre outwards and a new cell wall is laid down.

Significance of mitosis

→ Mitosis produces two cells which have the same number of chromosomes as the parent cell and each chromosome is an exact replica of one of the originals, i.e. they have identical DNA.

→ The division allows the production of cells that are genetically identical to the parent and so gives genetic stability.

→ By producing new cells, mitosis leads to growth of an organism and also allows for repair of tissues and the replacement of dead cells.

→ An additional function of mitosis is to provide for asexual reproduction, e.g. in plant cuttings, bacteria.

Checkpoint 3

Which cell organelles are concerned with spindle formation?

Checkpoint 4

In what situation might it be a disadvantage for the offspring to be identical to the parent?

Test yourself

A question on mitosis may provide you with drawings of unnamed stages in the in correct order. You may be required to name the stages and give the correct order.
Draw diagrams of the stages and glue them on to cards. On separate pieces of paper write out the names of the stages. Then lay out the cards in the correct order and label the stages!

Exam practice   answers: page 34

1. Complete the labels for the diagrams below which show two phases of mitosis.

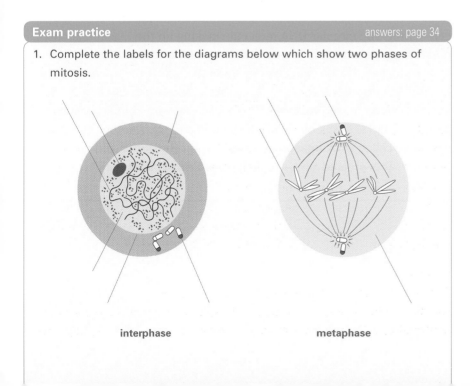

interphase   metaphase

Gene technology

Gene technology or genetic engineering has many commercial applications. Products such as insulin, enzymes and hormones can be produced on a large scale. In the near future it is hoped that the manipulation of genes will lead to the ability to treat, or even cure, genetic disorders.

Recombinant DNA technology

Genetic engineering involves the introduction of engineered DNA into cells in such a way that it will replicate and be passed on to progeny cells. One important application is the introduction of DNA from various organisms into bacterial cells, which then produce a required product. Recombinant DNA is formed when a piece of 'foreign' DNA is incorporated into the circular DNA (plasmid) from a bacterium.

To explain the principles of using gene technology to produce useful molecules on a large scale the production of **insulin** is described.

Isolating the gene from a donor DNA molecule

Insulin is a small protein. The gene coding for insulin is located using a gene probe and then isolated from the rest of the DNA in a human cell. An enzyme called **restriction endonuclease** is used to cut the DNA into small pieces allowing individual genes to be isolated. The enzyme cuts DNA between specific base sequences which the enzyme recognises. Most restriction enzymes split the two strands in a staggered sequence. The unpaired bases at the cut form **sticky ends**.

Inserting the gene into a vector

To get the gene into a bacterium a go-between or **vector** is used. The vector, in this instance, is a **plasmid**, a small, circular piece of DNA found in bacteria. In the case of insulin the bacteria used are called *Ecoli*. To obtain the plasmids the bacteria containing them are treated to dissolve their cell walls and the plasmids are separated from the cell debris. The circular DNA molecule making up the plasmid is cut open using the same enzyme, **restriction endonuclease**. The enzyme makes staggered cuts, called **sticky ends** that allow the donor DNA to be spliced into the vector DNA. This takes place when the donor and vector DNA are mixed together. The sticky ends are complementary and the C and G bases on their sticky ends pair up. Another group of enzymes, called **DNA ligases**, are used to join the donor and vector DNA together. This created DNA is called **recombinant DNA**.

Checkpoint 1

Why are prokaryotes more appropriate for recombinant DNA technology than eukaryotes?

Checkpoint 2

Name the specific biological molecule that forms the plasmid.

The jargon

The *sticky ends* are the two ends of the foreign DNA segment, which have a short row of unpaired bases that match the complementary bases at the two ends of the opened-up plasmid.

Checkpoint 3

Name the enzyme that splits the DNA molecule.

The jargon

A *plasmid* is a small circular loop of DNA found in bacterial cells. Bacterial plasmids are widely used as vectors.

Marker genes

Marker genes that confer antibiotic resistance are included in the recombinant DNA forming the plasmids. When the bacteria and plasmids are mixed together, only a small proportion of the bacteria take up the plasmids. If the bacteria are then cultured on a growth medium treated with antibiotic, only those containing the plasmid with the marker will survive. Using the bacteria that have taken up a piece of foreign DNA successfully, the foreign DNA replicates along with the rest of the plasmid every time the bacterial cell divides. **Cloning**, of the recombinant containing bacteria, results in the production of multiple copies of the recombinant gene. That is, the bacteria divide repeatedly and give rise to a large population of bacterial cells all of which contain replicas of the foreign DNA.

Manufacture and advantages

The genetically modified bacteria are cultured on a large scale using a fermenter and produce insulin which is extracted and purified. Thus, human insulin can be used instead of extracting insulin from animals. Also large quantities of product are produced quickly and relatively cheaply. Locating the correct piece of DNA is not easy as the cell contains two copies only of the DNA. However, the cell may contain large numbers of molecules of mRNA that has been transcribed from it. This is especially true if cells can be used which are known to synthesize and secrete the required product.

The functional mRNA will be present in large quantities in the cytoplasm.
→ This mRNA can be extracted.
→ The addition of an enzyme made from viruses, **reverse transcriptase**, is used to make a DNA copy of the mRNA. This single strand of DNA is called copy DNA or cDNA. Many copies of cDNA are made.
→ The addition of **DNA polymerase** converts this to a double strand for insertion into a plasmid as described above.

Action point

Draw a flowchart showing how a plasmid vector may be used in recombinant DNA technology.

Links

See pages 172 and 173 for further applications.

Checkpoint 4

E. coli is a bacterium used to clone genes. It is widely distributed and capable of exchanging genetic material with other types of bacteria. Suggest why some scientists think that the production of recombinant DNA could be dangerous.

Action point

Make a list and give definitions for the words in red on page 30.

Exam practice   answers: page 34

1 Describe how recombinant DNA is produced during the manufacture of insulin and outline the problems encountered with the use of DNA technology.

Answers
Molecules and cells

Molecules 1

Checkpoints

1 Five.
2 In α glucose the OH group on carbon 1 points downwards whereas the H group points upwards (in β glucose the H group points upwards and the OH group points upwards).
3 Condensation reaction.
4 β glucose.

Exam practice

1 (a) Plant cell wall.
  (b) (i) Glucose (no need to specify α or β form).
     (ii) Hexose.
     (iii) Parallel chains; cross-linked with hydrogen bonds.
     (iv) Tensile strength (strong) or porous.

Molecules 2

Checkpoints

1 The phosphate group is ionized and therefore water-soluble at that end of the molecule (hydrophilic).
2 Amino acids dissolve in water but are insoluble in organic solvents.
3 By means of hydrogen bonds.
4 An increase in kinetic energy causes the breaking of hydrogen and ionic bonds.
5 Globular.

Exam practice

1 (a) (i) A: amino group.
       B: carboxyl group.
    (ii) H.
  (b) (i)

    (ii) box on diagram.
    (iii) condensation.

Nucleic acids

Checkpoints

1 TCGGATGCA.
2 Organic bases.
3 In DNA the sugar is deoxyribose whereas in RNA the sugar is ribose.
  In DNA the base thymine is present whereas in RNA it is replaced by uracil.
  DNA is double-stranded, RNA is single-stranded.
4 DNA.

Exam practice

1 (a) (i) hydrogen.
    (ii) adenine or guanine.
    (iii) nitrogen; phosphorus.

  (b) Thymine in DNA is replaced by uracil in mRNA.
    DNA is a double helix while mRNA is usually not.
    DNA is double-stranded while mRNA is single-stranded.
    In DNA the sugar is deoxyribose and in mRNA it is a ribose sugar, or reference to the sugar in RNA containing more oxygen.
    In DNA the bases are paired and in RNA they are not.
    In DNA the ratio of adenine + thymine/cytosine + uracil is about equal while in RNA the ratio of adenine + uracil/cytosine + guanine is more variable.

Replication

Checkpoints

1 To prevent contamination, so ^{15}N is not incorporated.
2 Meselson and Stahl.
3 25% ^{15}N and 75% ^{14}N.
4 16.

Exam practice

1 (a) 50% light 50% intermediate.
  (b) When intermediate divides the heavy DNA strand would give intermediate DNA, the light strand would give light only. The intermediate DNA would give equal amounts of light again and the light form would produce only light.
  (c) DNA polymerase.
  (d) Multiplication of small traces of DNA to produce sufficient amounts for analysis in criminal investigations.

Protein synthesis

Checkpoints

1 672/3 = 224.
2 Ribosome or rough ER.
3 mRNA.
4 tRNA.
5 DNA.
6 tRNA.
7 mRNA.
8 There are four different bases in DNA but there are over 20 different amino acids. Having three bases for each amino acid would give a permutation of 64 codes, more than enough to make 20 amino acids. Amino acids are in fact coded by more than one DNA base triplet. Also there are some codes that do not code for amino acids at all, known as 'stop' and 'start' codes.

Exam practice

1 Your answer should include a detailed description of protein synthesis (see text) and should include diagrams.

Enzymes 1

Checkpoints

1 The number of substrate molecules which an enzyme can act upon in a given time is called its 'turnover number'. Catalase has a turnover number of several million, the highest known turnover number.
2 The site on the enzyme molecule where the substrate molecule is held and where the reaction occurs.
3 With increased temperature there is increased kinetic energy of the molecules, i.e. both substrate and enzyme molecules move about more rapidly and therefore the chance of collisions is increased. However, when a temperature of about 45 °C is reached, denaturation occurs (hydrogen bonds are broken as they are individually weak) and the shape of the protein molecule is irreversibly altered. At low temperatures, e.g. 0 °C, there is little movement of molecules and enzymes are said to be inactive. However, since no hydrogen bonds are broken, the shape remains unaltered and so a rise in temperature causes an increase in the rate of the reaction as the enzyme molecule has not been damaged.
4 No effect.
5 With immobilized enzymes:
- the product is not contaminated with enzyme
- enzyme is protected against changes in pH/temperature
- several enzymes with differing pH/temperature optima can be used together (a wider range of conditions can be tolerated).
6 Enzymes are more efficient, i.e. have a higher turnover number and are extremely specific. However, they are denatured by heat and affected by pH.

Exam practice

1 Your essay should cover all the following points in a logical sequence.
- Most reactions require energy to induce them to occur.
- This increases the activity of molecules, causing reactants to collide (or enhances break-up).
- The combination of reactants with enzymes reduces the required activation energy.
- This allows reactions to occur at physiological (body) temperatures.
- Enzymes are proteins.
- They are held in tertiary structure (or globular) by intermolecular bonds (hydrogen, disulphide, etc.) or the shape is held by bonding.
- This produces a specifically shaped active site.
- This allows the formation of an enzyme–substrate complex.
- You should provide a diagram of the lock and key mechanism.

Enzymes 2

Checkpoints

1 The rate of formation of product would increase. This is because the chance of an enzyme–substrate molecule contact is now greater than the chance of an enzyme–inhibitor contact.

2 A competitive inhibitor attaches to the active site whereas a non-competitive inhibitor attaches at a point other than the active site.
3 Competitive inhibition.
4 Non-reversible inhibition leaves the enzyme permanently damaged whereas if the reversible inhibitor is removed the enzyme can work normally once more.

Exam practice

1 (a) (i) curve A.
- The concentration of substrate limits the rate at low concentrations **or** at low concentrations there is more substrate for enzymes to work on.
- At high concentrations the enzyme is the limiting factor because the enzyme is working as fast as possible **or** there are not enough enzyme molecules present to combine with all the substrate molecules.
(ii) With the competitive inhibitor (curve B) there is an increased chance of the enzyme meeting the substrate. With non-competitive inhibitor (curve C) the enzyme is permanently inactivated.
(b) (i) The active site changes shape to accommodate the substrate.
(ii) With the lock and key theory it is not possible to alter the shape of the active site whereas the induced fit theory means the active site can change shape.
2 (a) Hydrogen/sulphur bonds are broken.
This alters the tertiary structure of enzymes **or** enzymes are denatured.
This changes the shape of the active site so that the substrate no longer fits.
(b) Using figures from the table, above 40 °C the rate of reaction is higher in immobilized enzymes.
(c) Trapping **or** encapsulation **or** bonding **or** adsorption.

Cell structure and cell membrane

Checkpoints

1 Prokaryote: no nuclear membrane, circular strands of DNA, no membrane-bound organelles, no meiosis/mitosis. Eukaryote: membrane-bound nucleus, chloroplasts and mitochondria may be present, meiosis and mitosis occur.
2 Plasma membrane/cell membrane.
3 Cell wall.
4 Extrinsic proteins are found in or on only one of the layers of the membrane whereas intrinsic extend across both layers.
5 As opposed to static, dynamic suggests movement, i.e. the protein floats about, although some proteins are anchored to organelles within the cell. The lipid component also moves.
6 Fat-soluble molecules.
7 In between the phospholipid molecules.

Exam practice

1 (a)
- The membrane consists mainly of phospholipid molecules which have a hydrophilic head and a hydrophobic tail.
- Give a description or diagram showing glycerol joined to two fatty acids and one phosphate molecule.
- The glycerol and phosphate is hydrophilic.
- The fatty acid part is hydrophobic.
- Since both sides of the plasma membrane are in contact with aqueous solutions the hydrophilic heads are orientated both inwards and outwards, forming a bilayer with the tails in the centre.

(b)
- Membrane phospholipid molecules have weak intermolecular forces and can move freely relative to one another.
- There are 'floating islands' of protein held in a membrane.
- Some (intrinsic) proteins pass right through the bilayer.
- Some (extrinsic) proteins float in the outer half of the bilayer or are attached to its surface.

Cell organelles

Checkpoints

1 To allow mRNA to pass through from the nucleus to the cytoplasm.
2 Chloroplast.
3 Mitochondrion.
4 Free ribosomes produce proteins which are retained in the cell, whereas ribosomes attached to ER produce proteins which are transported or secreted outside the cell.

Exam practice

1 (a) A: plasma membrane **or** plasmalemma **or** cell membrane.
   B: phagocytic vesicle **or** food vacuole.
   C: lysosomes **or** Golgi vesicle.
   D: Golgi apparatus.
   (b) Label ribosome on rough ER.

Chromosome structure and mitosis

Checkpoints

1 Centromere.
2 Interphase.
3 Centrioles.
4 If the environment changes there will be no variants that might be more suited to the environment.

Exam practice

1 This answer provides labelling only. You should refer to your notes for descriptions of phases.

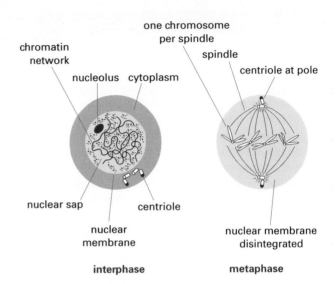

interphase   metaphase

Gene technology

Checkpoints

1 All the genes of prokaryotes are expressed in the next generation since the DNA is single-stranded.
2 DNA.
3 Restriction endonuclease.
4 Introduced DNA could spread among bacterial, plant and animal populations with unpredictable results.

Exam practice

1 Your answer should include the following points using diagrams to illustrate your answer.
- Insulin gene/DNA sequence must be identified
- Use gene probes
- Gene isolated using restriction enzyme
- Or made from RNA using reverse transcriptase
- DNA polymerase to produce double strand from cDNA
- Suitable vector used, e.g. plasmid
- Vector DNA cut using same restriction enzyme
- Complementary base pairing
- Original joined to vector DNA using ligase
- Reference to sticky ends
- Reference to antibiotic marker
- Problems – technically complex, expensive to develop, identifying useful genes, eukaryotic may not be expressed in prokaryotes, transfer of antibiotic resistant genes, possible transfer to oncogenes.

Revision checklist
Molecules and cells

By the end of this chapter you should be able to:

1	Relate the structure of carbohydrates, fats and proteins to their functions.	Confident	Not confident **Revise** pages 10–12
2	Describe using diagrams, the formation of bonds in the synthesis of disaccharides, fats and dipeptides.	Confident	Not confident **Revise** pages 10, 11
3	Describe the four levels of protein structure.	Confident	Not confident **Revise** page 12
4	Identify the components of a nucleotide and describe how hydrogen bonding between complementary base pairs leads to the formation of a DNA molecule.	Confident	Not confident **Revise** page 14
5	Outline with the aid of diagrams how DNA replicates semi-conservatively.	Confident	Not confident **Revise** page 17
6	Distinguish between the three forms of RNA.	Confident	Not confident **Revise** page 15
7	Describe the functions of the various nucleic acids in protein synthesis.	Confident	Not confident **Revise** pages 14, 15
8	Explain the mechanisms of action of enzyme molecules with reference to specificity, active site, enzyme-substrate complex.	Confident	Not confident **Revise** page 20
9	Describe and explain the effect of temperature, pH, enzyme concentration and substrate concentration in enzyme cells.	Confident	Not confident **Revise** page 20
10	Explain the effects of competitive and non-competitive inhibition on the rate of enzyme controlled reactions.	Confident	Not confident **Revise** page 22
11	Tabulate differences between prokaryote and eukaryote cells.	Confident	Not confident **Revise** page 24
12	Recognise the range of organelles as seen under the electron microscope and describe their functions.	Confident	Not confident **Revise** pages 26, 27
13	Describe the fluid mosaic model of membrane structure and give the functions of the components.	Confident	Not confident **Revise** page 26
14	Describe the main stages of mitosis and explain the significance of this form of cell division.	Confident	Not confident **Revise** page 28
15	Explain the steps involved in the production of insulin by recombinant DNA technology.	Confident	Not confident **Revise** page 30

Exchange and transport

All living organisms exchange gases with the environment. The raw materials for processes like respiration and photosynthesis must be transported to the cells which need them and waste products must be carried away. Soluble food absorbed by the gut wall of animals must be transported around the body, and the products of photosynthesis must be taken to all parts of the plant. Materials also need to pass in and out of cells as they are required and produced by the various cell organelles.

Exam themes

- How membranes are involved in the movement of materials in and out of cells
- Methods of movement of molecules
- The structure of the human gut and associated organs
- The structure of the ileum and its functions in digestion and absorption
- The opening and closing mechanism of stomata
- The rate of respiration and its nervous control
- The control of ventilation
- The effect of activity on the respiratory cycle
- Functions of blood in the transport of gases
- The significance of the dissociation curves of adult oxyhaemoglobin at different carbon dioxide levels
- The relationship between the structure and function of arteries, veins and capillaries
- The structure of red and white blood cells and the differences between blood, tissue fluid and lymph
- The structure and function of the mammalian heart
- Heart rate and the control of heartbeat
- The structure of xylem and phloem in relation to their transport functions
- The pathway of water transport
- Factors affecting the rate of transpiration
- Adaptations of plants to reduce water loss
- The mass flow theory of translocation in the phloem
- Evidence for transport in phloem

Topic checklist

	Edexcel		AQA		OCR		WJEC	
	AS	A2	AS	A2	AS	A2	AS	A2
Movement of molecules	O		O		O		O	
Human digestion			O				O	
Gas exchange in plants and simple organisms	O		O		O		O	
Gas exchange in mammals	O	●	O		O		O	
Transport in mammals: blood	O		O		O		O	
Transport in mammals: vessels	O		O		O		O	
Transport in mammals: heart	O	●	O		O		O	
Transport in plants	O		O		O		O	
Transport of water through the plant	O		O		O		O	
Translocation in plants					O		O	

Movement of molecules

The organelles and structures within a cell require a variety of materials in order to carry out their functions. In turn they form products, some useful and some waste. To understand plants and animals as whole organisms you first need to look at how materials enter and leave cells.

Checkpoint 1

Suggest how an increase in temperature would affect the rate of diffusion.

Examiner's secrets

Using expressions such as 'short diffusion path' in the correct context will impress.

Diffusion

Diffusion is the movement of molecules or ions from a region where they are in high concentration to a region of lower concentration. Ions and molecules are always in a state of random movement, but if they are highly concentrated in one area there will be a net movement away from that area until equilibrium is reached or until there is a uniform distribution. The rate of diffusion depends on:

→ the concentration gradient – a steeper gradient means a faster rate
→ the distance over which it takes place – the shorter the distance the greater the rate
→ the surface area – the larger the surface area the greater the rate
→ temperature – important in plants and cold-blooded animals

Facilitated diffusion

Links

Remember the structure of the cell membrane given on pages 24 and 25.

Charged particles or ions and large molecules such as glucose do not readily pass through the cell membrane because they are relatively insoluble in lipid. In the cell membrane protein molecules span the membrane from one side to the other and help such particles to diffuse in or out of the cells. These proteins are of two types.

Checkpoint 2

Explain how the structure of the cell membrane determines the preferential uptake of lipid-soluble molecules compared to water-soluble molecules.

→ *Channel proteins*: these form water-filled pores in the membrane. As the channel is hydrophilic, water-soluble ions can pass through. The channels are selective and in this way the cell can control the entry and exit of molecules and ions.
→ *Carrier proteins*: these bind molecules to them and then change shape as a result of this binding in such a way that the molecules are released on the other side of the membrane. They conduct molecules in either direction and energy from ATP is not needed.

Checkpoint 3

What would happen to the appearance of cells placed in concentrated glucose solution?

Osmosis

Watch out!

Water potential is always expressed as a negative value.

In biological systems osmosis is a special form of diffusion which involves the movement of water molecules. Most cell membranes are permeable to water and certain solutes only. These membranes are termed partially permeable. The term 'water potential' (WP) is used by biologists to describe the tendency of water molecules to move from high concentrations of water molecules to lower concentrations, i.e. water will move from a region of high WP to one of lower WP. This is because where there is a high concentration of water molecules they have a greater potential energy. A higher WP implies a greater tendency to leave. The WP of pure water is zero, so all solutions have lower WPs than pure water and are given negative values. Water will diffuse from a region of less negative (higher) WP to one of more negative (lower) WP.

Checkpoint 4

Cell A has a WP of −10 kPa and cell B has a WP of −20 kPa. In which direction will water move?

If a plant cell is surrounded by pure water the water will flow by osmosis across the cell membrane and cytoplasm and into the vacuole. The vacuole swells causing a pressure against the cell wall. The internal hydrostatic pressure increases until the cell can take in no more water and the cell is said to be turgid.

Active transport

Unlike the processes described so far, active transport is an energy-requiring process in which ions and molecules are moved across membranes against a concentration gradient.

→ Ions and molecules can move in the opposite direction to that occurring in diffusion, i.e. against a concentration gradient.
→ The energy is supplied by ATP, and anything which affects the respiratory process will affect active transport.
→ The process occurs through the carrier proteins that span the membrane.
→ The proteins accept the molecule and then the molecule enters the cell by a change in shape of the carrier molecule.

Exam practice   answers: page 58

1 (a) Active transport, facilitated diffusion and osmosis are three of the ways in which substances move into cells. Complete the table by ticking the statements that are true for the particular method of transport.

	Method of transport into cell		
Statement	Active transport	Facilitated diffusion	Osmosis
Requires energy from ATP			
Involves protein carrier molecules			
Involves movement of molecules from where they are in a high concentration to where they are in a lower concentration			

(b) A turgid plant cell was placed in a solution of sucrose. The diagram below shows the appearance of the cell after 1 hour.

(i) From the diagram, what is the evidence which shows that the water potential of the cell sap must be higher than that of the sucrose solution?

(ii) Explain why the water potential at point X is equal to that at point Y.

Checkpoint 5

What is meant by the terms 'hypertonic', 'hypotonic' and 'isotonic'?

The jargon

Turgor (being turgid) plays an important role particularly in young and herbaceous plants. It supports them and maintains their shape and form.

Action point

Draw annotated diagrams to show the effect of immersing a partially turgid plant cell in
(a) pure water
(b) a solution with a solute concentration exceeding that of the cell sap in the vacuole

Links

See mineral salt uptake by roots on page 55.

Checkpoint 6

How will an increase in temperature affect active transport?

Checkpoint 7

List three processes that involve active transport in plants and animals.

Action point

Draw diagrams to show how carrier proteins change shape during active transport.

Don't forget

Any practical work connected with water potential and plasmolysis of cells.

Human digestion

A heterotrophic organism cannot synthesize its major food requirements and so depends on a source of complex organic molecules – in other words, it needs to eat! The food is required as a source of energy for activities such as movement, and for the synthesis of body tissues. The organic molecules must be broken down by digestion and absorbed into the body tissues from the digestive system before being used in the body cells. You need to study what happens in each part of the alimentary canal.

Digestion

The absorption of nutrients by the gut epithelial cells is only possible if the large molecules, carbohydrates, fats and proteins, are first broken down or digested into smaller products by means of enzymes. At the completion of digestion carbohydrates are broken down to monosaccharides, proteins to amino acids, and fats to fatty acids and glycerol.

Mouth

Digestion starts in the mouth, with chewing of the food by the teeth. While chewing is in progress, the food is also mixed with saliva from the salivary glands. Saliva is a watery secretion containing mucus and **salivary amylase**, together with some mineral ions. It is important for lubricating the food before it is swallowed. Amylase is an enzyme that breaks down starch to maltose. The food is formed into a bolus, which passes into the oesophagus where it is moved down to the stomach by peristaltic contractions of the longitudinal and circular muscles.

Stomach

The presence of food in the mouth stimulates the secretion of the **gastric** juice from the gastric glands in the stomach wall. The stomach is a muscular sac that contracts rhythmically and mixes up the food with the gastric juice. The gastric glands are simple tubular glands with peptic cells, oxyntic cells and mucus-secreting cells.

→ *Peptic cells* secrete the protein-digesting enzyme pepsin as an inactive precursor, pepsinogen. This prevents the enzyme from damaging the stomach tissues before it is released. When it is activated, it is prevented from damaging the stomach wall by the secretion of mucus.

→ *Oxyntic cells* secrete hydrochloric acid, which makes the stomach contents acid; the acid conditions kill off many pathogenic bacteria as well as activating the protein-digesting enzymes; pepsinogen becomes pepsin and begins the breakdown of proteins to polypeptides; pro-rennin becomes rennin, which coagulates the soluble protein in milk, so this enzyme is especially important in young mammals.

Examiner's secrets

You do not have to learn about all the enzymes involved in digestion.

Checkpoint 1

Waves of muscular contraction (peristalsis) move across the stomach wall after food arrives there. What part does peristalsis play in digestion?

Checkpoint 2

What is the difference between an exocrine and an endocrine gland?

Checkpoint 3

Would you classify the pancreas as an endocrine gland, exocrine or both?

Checkpoint 4

Why is it essential that digestive enzymes such as proteases be secreted in an inactive state?

The jargon

A *precursor* is an inactive enzyme. The precursor is activated only when food is present.

Links

Remember the structure of carbohydrates, fats and proteins on pages 10–11.

Small intestine

The small intestine in humans is divided into two main regions, the duodenum and the ileum. The **duodenum** comprises the first 20 cm of the small intestine and receives secretions from both the liver and the pancreas.

→ Bile is produced in the liver and stored in the gall bladder, from where it passes into the duodenum via the bile duct. It contains no enzymes but the bile salts are important in emulsifying the lipids present in the food. Emulsification is achieved by lowering the surface tension of the lipids, causing large globules to break up into tiny droplets. This enables the action of the enzyme lipase to be more efficient as the lipid droplets now have a much larger surface area. Bile also helps to neutralize the acidity of the food as it comes from the stomach.

→ **Pancreatic** juice is secreted from the exocrine glands in the pancreas and enters the duodenum through the pancreatic duct.
  It contains a number of different enzymes:
  → *pancreatic amylase*, which breaks down any remaining starch to maltose;
  → *pancreatic lipase*, which splits lipids into fatty acids and glycerol;
  → *trypsin* (secreted as the inactive precursor trypsinogen), which continues the breakdown of proteins.

In the walls of the duodenum are Brunner's glands, which secrete an alkaline juice and mucus. The alkaline juice helps to keep the contents of the small intestine at the correct pH for enzyme action, and the mucus is for lubrication and protection.

Enzymes are secreted from groups of cells at the bottom of the crypts of Lieberkuhn. Here maltase splits maltose into two glucose molecules and endopeptidases and exopeptidases complete the digestion of proteins to amino acids.

Absorption

Absorption follows digestion and takes place mainly in the small intestine. The surface available for absorption is greatly increased by the presence of villi.

→ Glucose and amino acids are absorbed across the epithelium of the villi by a combination of diffusion and active transport, and pass into the capillary network that supplies each villus.

→ Fatty acids and glycerol pass into the epithelial cells and recombine to form neutral fat which is then passed into the lymphatic system.

Checkpoint 5

What changes in pH occur in the various regions of the gut during the digestion of food?

Checkpoint 6

Name two essential constituents of food that are not digested.

Checkpoint 7

Name three products of the digestion of bread.

Test yourself

For the process of digestion complete a table with the following headings: organ of secretion, site of action, contents, effect.

The jargon

Peptidases are enzymes that digest proteins and there are two groups. *Endopeptidases* hydrolyze the peptide bonds between amino acids in the central region of the polypeptide. *Exopeptidases* hydrolyze the peptide bonds on the terminal amino acids, progressively reducing them to their individual amino acids.

Action point

Draw a detailed section through a villus to show regions of absorption.

Exam practice   answers: page 58

1 Describe and illustrate the structure and histology of the mammalian small intestine. Explain fully the ways in which the small intestine is adapted to carry out its functions.

Gas exchange in plants and simple organisms

Gas exchange involves the exchange of oxygen and carbon dioxide between the organism and its environment. It takes place at a respiratory surface by the process of diffusion. You should understand the role of stomata in the exchange of gases in plants.

The respiratory surface

In order to achieve the maximum rate of diffusion a respiratory surface must have the following features:

→ thin so that diffusion paths are short
→ permeable to the respiratory gases
→ moist to allow easier diffusion of gases
→ a sufficiently large surface area to satisfy the needs of the organism

Checkpoint 1

Why must all microscopic organisms live in water?

Single-celled organisms

In simple, single-celled organisms diffusion of gases occurs over the whole of the body surface. The surface area to volume ratio is large enough to satisfy the organism's needs, the diffusion paths are short and there is no need for a circulatory system.

Checkpoint 2

Suggest how organisms such as flatworms overcome their problem of being multicellular without needing a circulatory system.

Plants

The structure of the leaves of flowering plants (see diagram below) is related to their function of gaseous exchange. To enable gaseous exchange to take place efficiently:

Action point

Draw a labelled diagram of a transverse section of a leaf.

→ the lamina is thin so diffusion paths for gases are short
→ the spongy mesophyll tissue allows for the circulation of gases
→ the plant tissues are permeated by air spaces
→ the stomatal pores permit gas exchange
→ leaves have a large surface area to volume ratio

Checkpoint 3

Explain why in the daylight plants take in carbon dioxide and give out oxygen and so appear not to respire.

Labels: lamina, vein, cuticle, upper epidermis, palisade mesophyll, spongy mesophyll, lower epidermis, cuticle, stoma, guard cell, air chamber

Action point

Rephrase the bullet lists as an essay-style paragraph to describe how the leaf is adapted to its function of gaseous exchange.

Stomata

These are pores in the epidermis, each bordered by two guard cells. The guard cells differ from the other epidermal cells in that they possess chloroplasts. The inner wall of each guard cell is thicker than the outer wall.

Gases diffuse through the stomata along the concentration gradient. Once inside the leaf the gases in the sub-stomatal air chambers diffuse through the intercellular spaces between the mesophyll cells and diffuse into the cells. The direction of diffusion depends on the environmental conditions and the requirements of the plants. It is the net exchange of carbon dioxide and oxygen in relation to respiration and photosynthesis which matters.

Because the inside of the leaf needs to be open to the atmosphere for the exchange of gases it is inevitable that water is lost from the leaves by the process of transpiration. In conditions of water deficit the stomata close.

The opening and closing mechanism of stomata

During the day the mechanism for stomatal opening occurs as follows.

→ A potassium ion (K^+) pump in the cell membranes actively transports K^+ ions into the guard cells.

→ As photosynthesis occurs the carbon dioxide concentration falls.

→ The pH rises and an enzyme catalyzes the conversion of starch to malate.

→ K^+ and malate ions accumulate in guard cells.

→ The water potential is lowered and water flows in by osmosis.

→ Guard cells become turgid and curve apart more because their outer walls are thinner than the inner walls, so the pore widens (see diagram below).

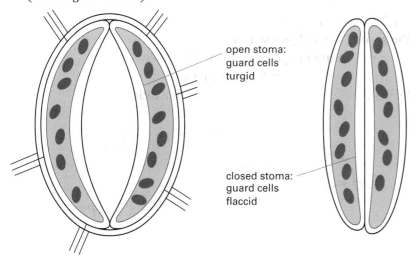

open stoma:
guard cells
turgid

closed stoma:
guard cells
flaccid

Checkpoint 4

Apart from xylem and phloem cells, which cells of the leaf do not contain chloroplasts?

Checkpoint 5

Which three environmental conditions affect the rate of diffusion?

Links

Transpiration is discussed on page 53.

Checkpoint 6

Which organelle must be in abundant supply for the K^+ ion pump to operate?

Links

Osmosis is discussed on pages 38 and 39.

Test yourself

Explain why stomata close at night.

Action point

Describe the mechanism for stomatal **closure**. It is the reverse of the opening mechanism described.

Exam practice   answers: page 56

1 Outline the functions of the stomata of a leaf. Give an illustrated account of the structure and function of the stomata, describing the theories which have been put forward to explain their opening and closing.

Gas exchange in mammals

Mammals, being relatively large, have a small surface area to volume ratio and often have a higher rate of metabolism than unicellular organisms. They therefore require more oxygen and produce more carbon dioxide. Gases cannot readily penetrate their surface so they, like other multicellular animals, have evolved special gaseous exchange mechanisms.

Overcoming problems associated with increase in size

An increase in size and complexity brings about certain problems:

→ the surface is some distance from most of the cells
→ there is an increase in metabolic rate
→ there is a decrease in surface area to volume ratio

Terrestrial vertebrates have evolved four main ways to overcome these.

→ *Lungs*: a compact respiratory surface but with a large surface area.
→ *A ventilation mechanism*: moves a fresh supply of air over the respiratory surface and maintains diffusion gradients.
→ *An internal transport system*: a blood circulation system moves gases between respiring cells and the respiratory surface.
→ *A respiratory pigment* in the blood increases its oxygen-carrying capacity.

Structure of the respiratory system

The lungs are enclosed within an airtight compartment, the **thorax**, at the base of which is a dome-shaped sheet of muscle called the **diaphragm**. Air is drawn into the lungs via the trachea. The lungs consist of a branching network of tubes called bronchioles arising from a pair of bronchi (see diagram below).

bronchiole

ring of cartilage

bronchus

alveolus

Ventilation mechanism

Breathing in is called inspiration and breathing out is called expiration. Inspiration is brought about as follows. The muscles between the ribs, called the external **intercostal** muscles, contract. At the same time the diaphragm contracts, and flattens. The ribcage moves up and out. This increases the volume of the thorax. The pressure in the thorax falls and air rushes in from the exterior.

Examiner's secrets

Exam questions are often based on the theme of how specific respiratory surfaces are adapted for efficient gaseous exchange.

Checkpoint 1

Name the respiratory pigment found in blood.

Checkpoint 2

How is the trachea protected against closure?

Watch out!

Be prepared to compare drawings of normal and diseased lung tissue.

Test yourself

Describe the process of expiration.

Gaseous exchange in the alveolus

In the alveoli there is a thin film of moisture in which the oxygen dissolves and then diffuses into the blood capillaries. Carbon dioxide diffuses from the blood into the alveoli (see diagram below).

Rate of respiration

Human lungs have a volume of about $5dm^3$, but at rest only about $0.45dm^3$ of this will be exchanged. This is called the tidal volume. With increasing activity both the frequency and depth of inspiration will increase. Vital capacity is the total volume of air that can be expired after a maximum inspiration. Even after maximum expiration about $1.5dm^3$ of air remains in the lungs. This is called the residual volume.

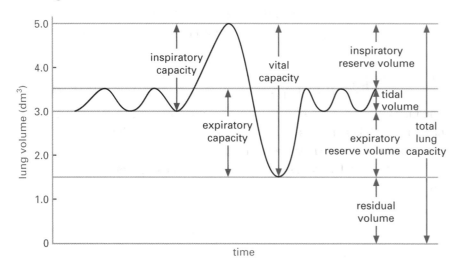

Nervous control of breathing

Breathing is controlled by a respiratory centre in a region of the hindbrain called the medulla oblongata. The respiratory centre itself is sensitive to pH. Also, chemoreceptors monitor blood pH and therefore detect changing carbon dioxide concentrations. The chemoreceptors are located in:

→ the carotid bodies in the carotid arteries of the neck

→ the aortic body in the wall of the aorta close to the heart

Links

The carriage of oxygen and carbon dioxide in the blood is dealt with on pages 46 and 47.

Checkpoint 3

List the basic requirements of a respiratory surface.

Checkpoint 4

Calculate your ventilation rate (breathing in and out counts as 1).

Checkpoint 5

State two effects on your breathing when taking exercise.

Checkpoint 6

What does a spirometer measure?

Action point

Study the graph opposite and state the volumes of gas exchanged for each of the terms.

Checkpoint 7

What is meant by the term 'residual volume'?

Exam practice   answers: page 58

1 Give an illustrated account of the structure of the mammalian respiratory system.

Transport in mammals: blood

In mammals a transport system provides the link between specialized areas for gas exchange and the cells which require oxygen and nutrients. As all cells are bathed in an aqueous medium, the delivery of materials to and from these cells is carried out largely in solution. The fluid in which the materials are dissolved or suspended is blood.

Composition of blood

Checkpoint 1

Red blood cells live for only three months. Where are they recycled?

Blood is a tissue made up of cells in a fluid plasma. *Plasma* is made up largely of 90% water, with soluble food molecules, waste products, hormones, plasma proteins, mineral ions and vitamins dissolved in it. The cellular elements (blood cells and cell fragments) are of three types:

→ **erythrocytes** or red blood corpuscles
→ **leucocytes** or white blood corpuscles
→ **thrombocytes** or platelets, which are fragments of cells

Checkpoint 2

Why are red blood cells biconcave in shape?

Functions of blood

Test yourself

Make up a table of the functions of the parts of the blood.

→ Plasma transports digested food products, hormones, proteins, fibrinogen, antibodies, etc. and also distributes heat.
→ Leucocytes are of two groups:
  → granulocytes are phagocytic, have granular cytoplasm, lobed nuclei and engulf bacteria;
  → agranulocytes produce antibodies and antitoxins, have clear cytoplasm and spherical nuclei.
→ Thrombocytes are involved in blood clotting.
→ Erythrocytes are filled with the pigment haemoglobin, are biconcave in shape and do not contain a nucleus. Their function is the carriage of oxygen, described below.

Test yourself

Draw labelled diagrams of red and white blood cells.

Oxygen carriage

An efficient respiratory pigment readily picks up oxygen at the respiratory surface and releases it on arrival at the tissues. Respiratory pigments have a high affinity for oxygen when the concentration is high but a low affinity when the concentration is low. When **haemoglobin** is exposed to a gradual increase in oxygen partial pressure it absorbs oxygen rapidly at first but more slowly as the partial pressure increases. This relationship is known as the oxygen dissociation curve (see graph on next page).

The jargon

Oxygen is measured by *partial pressure*, otherwise called *oxygen tension*. As oxygen makes up around 21% of the atmosphere the partial pressure of oxygen in the atmosphere is around 21 kPa.

The release of oxygen from haemoglobin is facilitated by the presence of carbon dioxide, a phenomenon known as the **Bohr effect**.

→ When the partial pressure of oxygen is high, as in the lung capillaries, oxygen combines with haemoglobin to form oxyhaemoglobin.
→ When the partial pressure of oxygen is low, as found in the respiring tissues, then oxygen dissociates from haemoglobin.
→ When the partial pressure of carbon dioxide is high, haemoglobin is less efficient at taking up oxygen and more efficient at releasing it.

Examiner's secrets

Answers to questions on the 'Bohr effect' are concerned with: the *uptake* of oxygen by haemoglobin even at low partial pressures and the *offloading* of oxygen by the presence of high levels of carbon dioxide at the tissues.

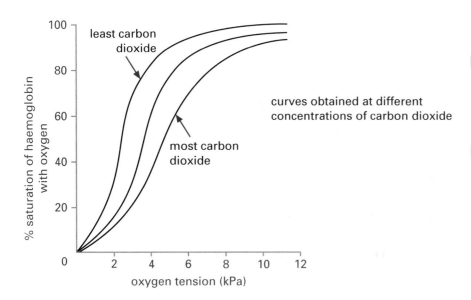

curves obtained at different concentrations of carbon dioxide

The jargon

The *Bohr effect* explains the way that oxygen moves from the red blood cell to the tissues. It is represented by a graph.

Checkpoint 3

What is the effect of the dissociation curve being displaced to the left?

Watch out!

Be prepared to explain the significance of the difference between the haemoglobin association curve of a fetus with that of an adult.

Carbon dioxide carriage

Carbon dioxide is transported in blood cells and plasma in three ways: 85% as hydrogencarbonate, 5% in solution in the plasma and 10% in combination with haemoglobin to form carbaminohaemoglobin.

→ Carbon dioxide diffuses into the red blood cell (RBC) and combines with water to form **carbonic acid**.

→ Carbonic acid dissociates into H^+ and HCO_3^- ions, the reaction being catalysed by **carbonic anhydrase**.

→ HCO_3^- ions diffuse out of the RBC into the plasma where they combine with Na^+ ions from the dissociation of sodium chloride to form sodium hydrogen carbonate.

→ H^+ ions encourage the **oxyhaemoglobin** to dissociate into oxygen and haemoglobin.

→ H^+ ions are buffered by their combination with haemoglobin and the formation of haemoglobinic acid (HHb).

→ The oxygen diffuses out of the RBC into the tissues.

→ To balance the outward movement of negatively charged ions, chloride ions diffuse in. This is known as the chloride shift, which maintains the electrochemical neutrality of the RBC.

Checkpoint 4

How does the haemoglobin of an animal with a high metabolic rate differ from that of human haemoglobin?

Checkpoint 5

What is the main purpose of the inward movement of chloride ions?

Test yourself

Draw a large diagram of a red blood cell and devise a flowchart showing the process known as the chloride shift.

Grade booster

Transport of CO_2 and O_2 are difficult concepts, so study them thoroughly.

Exam practice answers: page 59

1 The diagram below shows exchange taking place between a RBC and the fluid surrounding active muscle tissue.

   (a) Name X.

   (b) Name an organ in the body where this pattern of exchange would not occur.

   (c) Explain fully how the release of oxygen is brought about.

   (d) Suggest a reason for the movement of the chloride ions shown in the diagram.

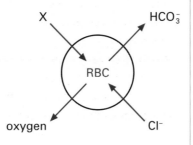

Transport in mammals: vessels

Grade booster

Discuss the possible effects of a high cholesterol level on the heart and circulatory system, with reference to HDLs and LDLs.

Action point

In table form compare open and closed blood systems.

Checkpoint 1

Why do arteries need a thick muscular wall?

Action point

Draw labelled diagrams of an artery, vein and capillary.

Action point

Make up a table of similarities and differences between arteries, veins and capillaries.

Checkpoint 2

A capillary network or bed slows down the flow of blood at the tissues. Why is this important?

Checkpoint 3

How does blood flow back to the heart?

A mammalian transport system incorporates a pump to sustain high pressure, valves to control the flow and vessels to distribute the blood. You should understand the structure and function of all these organs and vessels. You will also need to study the lymphatic system and its associated tissue fluid.

Open and closed systems

There are two types of blood system:

→ Animals such as insects have an **open blood system** whereby blood is pumped out of the heart at relatively low pressure into large spaces within the body cavity. There are no blood vessels. The blood bathes the tissues directly, exchange of materials takes place, and there is little control over the direction of circulation. Blood slowly returns to the heart.

→ In a **closed blood system** the blood circulates in blood vessels. Blood is pumped by a muscular heart at a high pressure resulting in a rapid flow rate. Organs are not in direct contact with the blood but are bathed by tissue fluid seeping out from thin-walled capillaries. The blood contains a blood pigment which carries oxygen.

Blood vessels and the circulation system

Blood is pumped by the heart into thick-walled vessels called **arteries**. These split up into smaller vessels called arterioles, from which the blood passes into the **capillaries**. The capillaries form a vast network which penetrates all the tissues and organs of the body. Blood from the capillaries collects into venules, which in turn empty blood into **veins**, which return it to the heart.

Structure of arteries and veins

The arteries and the veins have the same basic three-layered structure but the proportions of the different layers vary. In both:

→ the *innermost layer*, the endothelium, is one cell thick and provides a smooth lining with minimum resistance to the flow of blood

→ the *middle layer*, which is made up of elastic fibres and smooth muscle, is thicker in the arteries than in the veins

→ the *outer layer* is made up of collagen fibres that are resistant to overstretching

Veins have valves in them but these are not present in arteries apart from the aortic valves. The capillaries are thin-walled, consisting only of a layer of endothelium so that they are permeable to water and dissolved substances such as glucose. It is in the capillaries that the exchange of materials between the blood and the tissues takes place.

Intercellular fluid and lymph

When blood reaches the arterial end of a capillary it is under pressure because of the pumping action of the heart and the resistance to blood flow of the capillaries. The hydrostatic pressure forces the fluid part of the blood through the capillary walls into the spaces between the cells. This fluid is

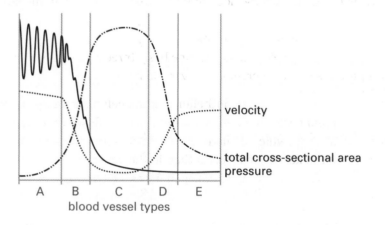

Action point

Explain the forces involved in the movement of water and solutes into and out of the capillaries.

Links

See osmosis on pages 38 and 39.

called tissue fluid. At the arterial end of the capillary bed it supplies the cells with oxygen and nutrients. In addition at the venous end, tissue fluid picks up carbon dioxide and other excretory substances. Blood pressure, diffusion and osmosis are the forces responsible for the movement of water and solutes into and out of the capillaries (see diagram above). Some of this fluid passes back into the capillaries, but some drains into the **lymphatic system** and is returned to the blood eventually via the thoracic duct, which empties into the subclavian vein.

Lymph is the tissue fluid that drains into blind-ending lymphatic capillaries among the tissues that join up to form larger vessels. It is moved by contractions of the muscles through which the vessels pass, being prevented from back-flow by numerous valves similar to those found in the veins. There are lymph glands and nodes associated with the lymph vessels, and these play an important role in the formation of lymphocytes and the prevention of infection.

Checkpoint 4

What are the two functions of lymph?

Exam practice

answers: page 59

1 The graph below shows the relationship between blood pressure, blood velocity and total cross-sectional area for the different types of blood vessels in the mammalian circulatory system.

Checkpoint 5

What components of the blood are normally not found in tissue fluid?

Watch out!

Pressure in blood vessels is highest in the arteries, the capillaries are the next highest in pressure and the veins have the least pressure in them. Can you explain why there are these differences in pressure?

(a) (i) Name the vessel types A–E.

　　(ii) Use the appropriate letter to indicate:

　　　　1 A type of vessel which contains the highest proportion of muscle fibre in its wall.

　　　　2 The vessels which contain most of the total blood volume.

(b) (i) Account for the variation in blood pressure observed in vessel type A.

　　(ii) Suggest a reason for the rapid fall in blood pressure in vessel type B.

(c) Comment on the physiological advantages of the relationship between blood velocity and cross-sectional area of vessel type C.

(d) Suggest two reasons for the increase in blood velocity in vessel types D and E when the blood pressure is so low.

Transport in mammals: heart

A pump to circulate blood is an essential feature of a circulatory system. The heart consists of thin-walled collection chambers and thick-walled pumping chambers which are partitioned into two, allowing the complete separation of oxygenated and deoxygenated blood. The heart is, in effect, two pumps side by side.

Checkpoint 1

Name the only artery to carry deoxygenated blood.

Checkpoint 2

Why is the left ventricle thicker-walled than the right ventricle?

Checkpoint 3

What is the function of the 'heart strings'?

The jargon

An *atrium* is also known as an *auricle*.

The jargon

Systole is a contraction phase of the cardiac cycle and *diastole* is the relaxation phase.

Examiner's secrets

The graphical analysis of pressure changes in the heart is a favourite exam question. Be prepared to describe the pressure changes involved in the movement of blood from one chamber of the heart to another together with the associated opening and closing of the valves.

Checkpoint 4

How is back-flow of blood into the atrium prevented?

Checkpoint 5

Describe the flow of blood through the left side of the heart.

Single and double circulations

Closed circulation systems are of two types:

→ Fish have a **single circulation**. The heart pumps deoxygenated blood to the gills, oxygenated blood is then carried to the tissues, from there deoxygenated blood returns to the heart. Blood goes once through the heart during each circuit of the body.

→ Mammals have a **double circulation**:

　⇢ The pulmonary circulation – the right side of the heart pumps deoxygenated blood to the lungs. Oxygenated blood then returns to the left side of the heart.

　⇢ Systemic circulation – the left side of the heart pumps the oxygenated blood to the tissues. Deoxygenated blood then returns to the right side of the heart.

In each circuit the blood passes through the heart twice, once through the right side and once through the left side.

The Heart

The four-chambered heart is situated in the thorax between the two lungs. It consists largely of **cardiac** muscle, a specialised tissue that is capable of rhythmical contraction and relaxation of its own accord throughout. The heart muscle is said to be '**myogenic**'.

The cardiac cycle

The cardiac cycle describes the sequence of events in one heartbeat. The pumping action of the heart consists of alternating contractions (**systole**) and relaxations (**diastole**).

There are three stages to the cardiac cycle:

→ Stage 1 – The right and left ventricles relax, the tricuspid and bicuspid valves open as the atria contract and blood flows into the ventricles.

→ Stage 2 – The atria relax and the right and left ventricles contract together forcing blood out of the heart into the pulmonary artery and the aorta as the semilunar valves are opened. The tricuspid and bicuspid valves are closed by the rise in ventricular pressure. The pulmonary artery carries deoxygenated blood to the lungs and the aorta carries oxygenated blood to the various parts of the body.

→ Stage 3 – The ventricles relax and the pressure in the ventricles falls. Blood under high pressure in the arteries causes the semilunar valves to shut, preventing blood from going back into the ventricles. Blood from the vena cavae and pulmonary veins enters the atria and the cycle starts again.

Pressure changes in the heart

→ The highest pressure occurs in the aorta/arteries that show a rhythmic rise and fall corresponding to ventricular contraction.

→ Friction with vessel walls causes a progressive drop in pressure. Arterioles have a large total surface area and a relatively narrow bore causing a substantial reduction from aortic pressure. Their pressure depends on whether they are dilated or contracted.

→ The extensive capillary beds have a large cross-sectional area. These beds create an even greater resistance to blood flow.

→ There is a relationship between pressure and speed and the pressure drops further due to leakage from capillaries into tissues.

→ The return flow to the heart is non-rhythmic and the pressure in the veins is low but can be increased by the massaging effect of muscles.

Control of heartbeat

→ The heart muscle is **myogenic**, i.e. the heartbeat is initiated from within the muscle itself and is not due to nervous stimulation.

→ In the wall of the right atrium is a region of specialized cardiac fibres called the **sinoatrial node** (SAN) which acts as a **pacemaker**.

→ A wave of electrical stimulation arises at this point and then spreads over the two atria causing them to contract more or less at the same time.

→ The electrical stimulation is prevented from spreading to the ventricles by a thin layer of connective tissue. This acts as a layer of insulation (it is important that the muscles of the ventricles do not start to contract until the muscles of the atria have finished contracting).

→ The stimulation reaches another specialized region of cardiac fibres, the **atrio-ventricular node** (AVN), which lies between the two atria and which passes on the excitation to specialized tissues in the ventricles.

→ From the AVN the excitation passes down the **bundle of His** to the apex. The bundle branches into **Purkinje fibres** in the ventricle walls which carry the wave of excitation upwards through the ventricle muscle.

→ The impulses cause the cardiac muscle in each ventricle to contract simultaneously from the apex upwards.

Grade booster

Study graphs showing volume and pressure changes in the cardiac cycle.

Action point

Draw an annotated diagram of the heart showing the control of the heartbeat.

The jargon

The *bundle of His* is continuous with the AVN and is a strand of modified cardiac muscle fibre that fans out over the walls of the ventricles.
The *Purkinje* tissue is a network of fibres just beneath the endothelial lining.

Grade booster

Research how the sympathetic and parasympathetic systems work on a negative feedback principle, involving two centres in the medulla of the brain.

Exam practice

answers: page 59

1 Complete the following by inserting the appropriate word or words.
The mammalian heart is made up of a special type of muscle called .... muscle. This muscle has the ability to contract and relax without any stimulation and is said to be .... The heartbeat is initiated in an area of the right atrium called the ..... . The wave of excitation passes across both atria until it reaches an area of tissue in the septum called the .... This in turn passes the wave to a group of fibres called the ...... which transfers the wave to the tip of the ventricles. This causes the ventricles to .... from the base upwards and forces blood to flow out of the heart through the aorta and ..... .

Examiner's secrets

In this exam question, abbreviations are not acceptable.

Test yourself

Mammals have a double circulation system, i.e. the blood moves through the heart twice during each cardiac cycle. Draw a section through the heart and from memory label the chambers and associated blood vessels. Use a coloured pen to show the circulation of blood through the heart.

Transport in plants

In plants the organs collecting the water, the roots, are some distance from the leaves, which require the water for the process of photosynthesis. A transport system through the stem is necessary to connect the two organs. You should study the distribution of xylem and phloem tissues in stems, roots and leaves, and the process by which water is lost through the leaves.

Watch out!

Don't confuse *xylem* and *phloem*.

Vascular tissues

The vascular tissues are made up of:

→ **xylem**, which transports water and mineral salts from the roots to the leaves
→ **phloem**, which mainly transports the soluble products of photosynthesis from the leaves to the other parts of the plant

Xylem distribution

The distribution of xylem tissue differs in primary stems, leaves and roots.

Checkpoint 1

State two differences between the arrangement of tissues in a stem and a root.

→ In *stems* it occurs as part of the peripheral vascular bundles. This gives flexible support, but also resistance to bending strain.
→ In *leaves* the arrangement of vascular tissues in the midrib and the network of veins gives flexible strength and resistance to tearing strains.
→ In *roots* the central arrangement is ideal for resistance to pull and so helps the anchorage of the plant.

Checkpoint 2

Describe how the arrangement of vascular tissue in the stem, root and leaf enables each to overcome the effect of the wind as a mechanical factor.

The different arrangements of xylem tissue are shown in the diagrams below.

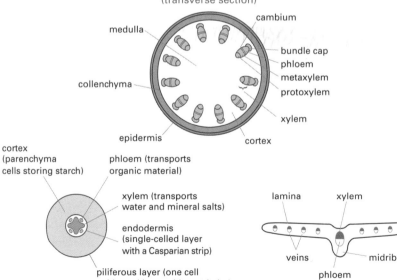

Low power map of tissues of young sunflower stem
(transverse section)

cambium
medulla
bundle cap
phloem
metaxylem
protoxylem
collenchyma
xylem
epidermis
cortex

cortex
(parenchyma
cells storing starch)

phloem (transports
organic material)

xylem (transports
water and mineral salts)

endodermis
(single-celled layer
with a Casparian strip)

piliferous layer (one cell
epidermal layer with root hairs)

TS dicotyledon root (*Ranunculus*)
primary structure

lamina   xylem

veins   midrib

phloem

TS dicotyledon leaf
(a typical mesophytic leaf)

Test yourself

From memory draw labelled diagrams of TS stem and root.

Transcription

The rate at which water is lost from plants is called the transpiration rate and is dependent on external factors such as temperature, humidity and air movements. A number of anatomical and morphological features of plants also influence the transpiration rate. These include the numbers and distribution of stomata, leaf area and thickness of the cuticle.

Some plants exhibit **xeromorphic** adaptations and are known as xerophytes. They need to prevent excessive water loss for one of a number of reasons. They may live in:

→ hot, dry desert regions
→ cold regions where the soil water is frozen for much of the year
→ exposed, windy situations

These xeromorphic adaptations may take a number of general forms:

→ *reducing transpiration rate* by having
  → rolled leaves
  → thick cuticle
  → depression of stomata
  → absence of leaves
  → reduction in surface area/volume ratio of leaves

→ *storing water* (succulents)
  → succulent leaves and stems
  → closing stomata during daylight

→ *resistance to dessication*
  → leaves reduced to spines with photosynthesis carried out by the stem
  → lignified cells supporting the leaf, preventing wilting and reduction in surface area for photosynthesis

Checkpoint 3

Using the term 'water potential' explain how more water is lost from a leaf on a dry, windy day than on a humid, still day.

The jargon

Anatomical features are internal.
Morphological features are external.

Checkpoint 4

What is meant by the term 'water stress'?

Checkpoint 5

Explain how sunken stomata can reduce transpiration.

Action point

Draw a labelled diagram of one example of a xerophytic adaptation as seen under the microscope.

Don't forget

The rate of transpiration is measured indirectly using an instrument called a potometer.

Exam practice   answers: page 59

1 *Hakea* is a plant which lives in the Australian desert. The diagrams below show sections of one of its leaves as seen through a microscope under low power and high power.

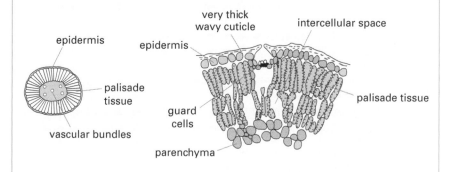

Suggest three ways in which the structure of its leaves helps *Hakea* survive in the hot, dry desert conditions.

Transport of water through the plant

The transport of water and minerals takes place in specialized vascular tissue, known as xylem. As you study this topic think of the forces of 'push' at the roots and 'pull' at the leaves.

Examiner's secrets

The distribution of vascular tissues differs in stems, roots and leaves. You should have some understanding of this from microscope studies of tissues and individual cells.

Checkpoint 1

How does lignification occur?

Test yourself

How is the structure of xylem related to its role of water transport?
Draw the four different types of xylem cells in LS and TS.
Consider the evidence supporting the view that xylem carries water up the stem.

Links

The effect of environmental conditions on transpiration is described on page 53.

The jargon

Plasmodesmata are tiny strands of cytoplasm which link the cytoplasm of one cell to that of the next.

Structure of xylem

Xylem is made up of four different types of cells: vessels, tracheids, fibres and xylem parenchyma. Vessels and tracheids are the cells involved in conduction. These are dead cells and they form a system of tubes through which water can travel. The cells are dead because lignin has been deposited on the cellulose cell walls rendering them impermeable to water and solutes. These cells also provide mechanical strength and support to the plant.

The movement of water through the plant

Water enters the plant mainly through the root hairs by osmosis, travels across the **cortex** and into the xylem. It then travels in the xylem up through the stem to the leaves, where most of it evaporates from the internal leaf surface and passes out, as water vapour, into the atmosphere.

The transpiration of water from the leaves draws water across the leaf from the xylem tissue along three pathways:

→ the **apoplast** – through the cell wall and intercellular spaces
→ the **symplast** – through the cytoplasm and plasmodesmata
→ the **vacuolar pathway** – from vacuole to vacuole

The transpiration of water causes a more negative water potential in the mesophyll of the leaf. As water in the xylem is of a higher (less negative) water potential, it moves up the xylem to replace the water lost. This movement of water is known as the transpiration stream. However, water also moves up the xylem by capillarity.

Capillarity has two components:

→ the water molecules have an attraction for each other (cohesion) so when one water molecule moves upwards, others move with it
→ the water molecules are attracted to the sides of the xylem vessels pulling the water upwards (adhesion). As xylem consists of very long, narrow vessels (or 'tubes'), considerable adhesive forces are developed, sufficient to support a considerable mass of water. The forces combine to maintain the column of water in the xylem.

This is known as the cohesion–tension theory and suggests an explanation for the rise of water in the xylem of all plants.

Water uptake by the roots

The large quantities of water lost through transpiration must be replaced from the soil. The region of greatest uptake is the root hair zone where the surface area of the root is enormously increased by the presence of root hairs on the cells of the piliferous layer.

Checkpoint 2

What is the other function of root hairs?

Checkpoint 3

Is osmosis an active or passive process?

Most of the water enters the root from the soil down a gradient of water potential. It then passes across the root cortex along two main pathways, the symplast and apoplast pathways, until it reaches the endodermis. Most of the water probably follows the latter pathway, which is the faster of the two. The cells of the endodermis have layers of suberin around them, forming a distinctive band known as the Casparian strip. This prevents the use of the apoplast pathway, so the water must pass through the cytoplasm of the cells in this layer, the only available route to the xylem. There is some evidence that salts may then be actively secreted into the vascular tissue from the endodermal cells. This makes the water potential in the xylem more negative, causing water to be drawn in from the endodermis, so promoting the movement of water into the xylem from the cortex. The water potential gradient produced creates a force known as the root pressure.

The uptake of minerals

Generally, minerals are taken up by the root hairs by active transport from the soil solution. Once absorbed the mineral ions may move along the apoplast pathway carried along in solution by the water being pulled up the plant in the transpiration stream. When minerals reach the endodermis the Casparian strip prevents further movement along the cell walls. The ions enter the cytoplasm of the cell, from where they diffuse or are actively transported into the xylem.

Exam practice   answers: page 60

1 The diagram below represents part of a transverse section across a young root. The relative distance across the cortex has been shortened for simplicity.

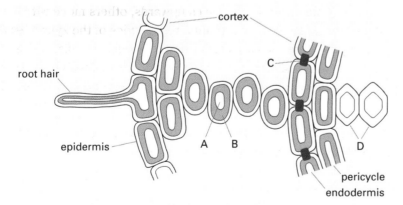

(a) (i) Name the features labelled A, B, C and D.
    (ii) State two advantages to the plant of root hairs.
    (iii) What is the function of feature C?
(b) It is believed that water can move across the cortex along the cellulose cell walls. Suggest how this movement is maintained.

The jargon

The *Casparian strip* is a thickening of a waxy material, suberin, on the radial walls of the endodermal cells. This impermeable layer diverts the water forcing it to take the symplast pathway.

Checkpoint 4

What happens if the stem of a plant is cut near the root during periods of rapid growth?

Action point

Construct a diagram of the plant showing the various pathways of water transport and indicate whether the processes involved are active or passive.

Checkpoint 5

What are the requirements of the active uptake of minerals?

Checkpoint 6

What is the effect of a respiratory inhibitor on mineral uptake?

Watch out!

Be prepared to draw different views of an endodermal cell.

Grade booster

Research how the Casparian strip may be involved in the selective absorption of minerals into the xylem.

Translocation in plants

The products of photosynthesis are transported in the phloem, away from the site of synthesis (the 'source') in the leaves, to all the other parts of the plant (the 'sink') where they are used for growth or storage. In plants the transport of the soluble organic materials, sucrose and amino acids is known as translocation.

Action point

Draw a sieve tube and companion cell.

Examiner's secrets

You should be able to list the experimental evidence that phloem is the transport tissue.

Examiner's secrets

Amino acids and sucrose (not glucose) are translocated in the phloem.

Structure of phloem

Phloem is a living tissue and consists of four types of cells:

→ sieve tubes
→ companion cells
→ phloem fibres
→ phloem parenchyma

The sieve tubes are the only components of phloem obviously adapted for the longitudinal flow of material. They are formed from cells called sieve elements placed end to end. The end walls do not break down but are perforated by pores. These areas are known as sieve plates. Cytoplasmic filaments containing phloem protein extend from one sieve cell to the next through the pores in the sieve plate. The sieve tubes do not possess a nucleus and most of the other cell organelles disintegrate. Each sieve tube element is closely associated with one or more companion cells, which have dense cytoplasm, large centrally placed nuclei, many mitochondria, and are connected to the sieve tube element by plasmodesmata.

Checkpoint 1

Which two food materials are transported in the phloem?

Checkpoint 2

Explain in terms of water potential the important differences between a 'source' and 'sink' region of the plant.

Checkpoint 3

What parts of the plant require sucrose?

Transport in the phloem

Experimental evidence suggests that the phloem is the tissue concerned with the translocation of organic substances.

→ Early evidence was obtained from **ringing experiments**.
→ More recently, the technique of **radioactive tracing** with labelled metabolites and direct sampling of the contents of individual sieve tubes, using the mouth parts of aphids as micro-pipettes, have been used. Radioisotope labelling is a technique whereby CO_2 labelled with radioactive carbon is supplied to an illuminated plant leaf. The radioactive carbon is fixed in the sugar produced in photosynthesis and its translocation to other parts of the plant can be traced using autoradiography. The technique shows that the sugar is transported in both an upward and downward direction since the radioactivity is shown in the aerial parts of the plant as well as the roots.

The observed rate of flow is much too rapid for diffusion to be the cause. The main theory put forward to explain the transport of organic solutes is known as the mass flow hypothesis (put forward in 1937 by Ernst Munch). This suggests that there is a passive mass flow of sugars from the phloem of the leaf where there is the highest concentration (the source) to other areas where there is a lower concentration (the sink).

- At the source, phloem companion cells actively take up sugars and pass it to the sieve tubes while the reverse process occurs at the sink. Transfer cells, which occur in the mesophyll, are found to be associated with the phloem tissue and are thought to be involved in this process.
- This causes a concentration gradient from source to sink.
- Water is drawn in by osmosis, setting up a hydrostatic pressure resulting in movement of sugar to the sink.
- At the sink the sugar concentration decreases, as it is used for metabolic processes, so water passes out into the tissues.
- The continual input of sugars and water at the top of the system and their removal at the bottom creates a pressure gradient that maintains the downward flow of fluid in the sieve tubes.

There are drawbacks to the theory.

- It does not explain the existence of the sieve plates, which seemingly act as a series of barriers impeding flow.
- Sugars and amino acids have been observed to move at different rates and in different directions in the same tissue.

Other hypotheses have been proposed, such as electro-osmosis, surface spreading and cytoplasmic streaming along the protein filaments.

Examiner's secrets

Check with your specification whether details of these theories are required.

Grade booster

Review the evidence for and against the mass flow hypothesis.

Exam practice   answers: page 60

1 (a) With the aid of labelled diagrams, describe the structures involved in the translocation of organic solutes between different parts in a flowering plant.

   (b) Explain one experiment which has been performed that has enhanced our knowledge of translocation in flowering plants.

2 (a) (i) Complete the table below which compares xylem and phloem.

Feature	Xylem	Phloem
Name of conducting cells		
Direction in which materials are transported		
One possible mechanism by which materials are transported		

      (ii) Briefly describe *one* way of demonstrating movement of fluid in phloem.

   (b) The table below shows the concentration of two chemicals to be found in the phloem and xylem sap of the white lupin, *Lupinus albus*.

Feature	Xylem (mg l^{-1})	Phloem (mg l^{-1})
Sucrose	not detected	154 000
Magnesium	27	85

   Suggest why the concentration of sucrose and magnesium differs in xylem and phloem.

Watch out!

Don't confuse transpiration and translocation.

Don't forget

Radioactive labelling techniques have been used to investigate the mechanism of translocation. A plant is able to use radioactive carbon dioxide in photosynthesis to produce labelled sugars.

Answers
Exchange and transport

Movement of molecules

Checkpoints

1 An increase in temperature results in an increase in rate since there is an increase in molecular energy and therefore movement.
2 Most of the membrane consists of lipophilic tails, which tend to repel ions and polar molecules. These can only diffuse across slowly whereas molecules with a high solubility in fat are taken up rapidly.
3 Shrink/shrivel.
4 A to B.
5 Solutions of equal concentration are called isotonic. A more dilute solution is called hypotonic. The more concentrated solution is called hypertonic.
6 Increases it since the rate of respiration is increased.
7 Plants – uptake of salts; animals – muscle contraction, nerve impulses, etc.

Exam practice

1 (a)

Active transport	Facilitated diffusion	Osmosis
✓		
✓	✓	
	✓	✓

(b) (i) Cell is plasmolyzed/cytoplasm shrunk from cell wall.
(ii) Cell wall is permeable (to sugar as well as water)/ same solution on both sides of cell wall.

Human digestion

Checkpoints

1 Thorough mixing of food and gastric juice.
2 An exocrine gland has a duct through which the secretions pass. In an endocrine gland blood flows through it and receives the secretions directly.
3 Both, since it produces insulin (endocrine) and digestive enzymes (exocrine).
4 To prevent autodigestion of the cells of the gastric glands in which they are formed.
5 Mouth – alkaline; stomach – very acid; duodenum – alkaline.
6 Water, minerals, vitamins.
7 Glucose, amino acids, fatty acids.

Exam practice

1
- General diagram of tissue layers showing outer layer of longitudinal and inner layer of circular muscle, inner mucosa, separated from muscle by submucosa.
- A labelled diagram of the gut wall showing villi, microvilli, epithelium, columnar cells, goblet cells, crypts, lacteals, capillaries, Brunner's glands (in duodenum only).
- Function of muscle, movement of food (peristalsis).
- Goblet cells producing mucus.
- Brunner's glands producing alkaline mucus to aid neutralization of acid from stomach.

- Cells in crypts of Lieberkuhn secrete enzymes.
- Length of gut important to allow time for enzyme action and absorption.
- The villi and microvilli increase surface area for absorption by epithelial cells.
- Further enhanced by microvilli on cells.
- In each villus capillaries 'pick up' simple sugars and amino acids which are carried away by the hepatic portal vein.
- Glycerol and fatty acids are removed by lacteals and are carried into the main lymphatic system.

Examiner's secrets

This question is specifically about the small intestine so don't be tempted to waste valuable time describing mouth and stomach. Be precise about which secretions are produced by the various glands.

Gas exchange in plants and simple organisms

Checkpoints

1 They would rapidly dessicate on land.
2 Being flattened they have a large surface area/also sluggish and have a low metabolic rate.
3 In daylight photosynthesis masks the respiratory production of carbon dioxide because the respiratory carbon dioxide is immediately reused in the chloroplasts.
4 Epidermal cells.
5 Humidity, wind, temperature.
6 Mitochondria.

Exam practice

1 The answer must include a diagram for full marks to be awarded. It should be clear and accurate with at least six labelled structures, including guard cells, thick inner wall and thin outer wall, nuclei, chloroplasts, cell vacuole.
The diagram should show a bending of the inner wall and opening of aperture in a turgid cell.
The mechanism is as described in the text.
The functions are to allow gaseous exchange and to prevent loss of water by closure in water-stressed plants.

Gas exchange in mammals

Checkpoints

1 Haemoglobin.
2 Rings of cartilage.
3 Thin, moist, permeable, large surface area.
4 Ventilation rate = number of breaths per minute (16) × tidal volume (16 × 0.45 = 7.2).
5 Increase in frequency and depth.
6 Allows the study of oxygen consumption in humans and can be used to record depth and frequency of breathing.
7 After maximum expiration this is the amount of air remaining in the lungs.

Exam practice

1 The answer must include a diagram for full marks to be awarded with at least six labelled structures. Include the following words in annotated labelling: pharynx, larynx, epiglottis, trachea, bronchi, cilia and mucus-secreting cells, bronchioles, alveoli, network of capillaries.

Transport in mammals: blood

Checkpoints

1 Liver.
2 Increased surface area and shorter diffusion path to carry oxygen.
3 It picks up oxygen more readily but is less ready to release it.
4 The haemoglobin has a low affinity for oxygen and therefore unloads it quickly to the tissues.
5 To ensure that the overall charge remains neutral.

Exam practice

1 (a) Carbon dioxide.
  (b) Lungs.
  (c) Chloride shift as described in the text.
  (d) It preserves the ionic balance on either side of the membrane.

Transport in mammals: vessels

Checkpoints

1 To withstand and maintain the pressure of blood when it is pumped away from the heart.
2 To slow blood flow to allow substances time to diffuse.
3 Contraction of skeletal muscles/inspiration.
4 Help prevent infection/lymphocytes produced in lymph nodes/tissue fluid bathes cells.
5 Tissue fluid lacks proteins (and blood cells).

Exam practice

1 (a) (i) A, artery; B, arterioles; C, capillaries; D, venule; E, veins.
     (ii) 1: B. 2: C.
  (b) (i) Contractions of left ventricle cause blood to surge into the artery; relaxation of the left ventricle results in a fall in pressure; the amplitude change is due to the elasticity of the vessels which reduces with distance from the heart.
     (ii) An increase in the total cross-sectional area of the vessels.
  (c) In the capillaries blood moves slowly because of the total cross-sectional area, maximizing time for the exchange of materials between the blood and tissues.
  (d) The massaging action of skeletal muscles; suction effects of the heart.

Transport in mammals: heart

Checkpoints

1 Pulmonary.
2 Pumps blood all round the body.
3 Prevent valves from turning inside out.
4 Tricuspid valves shut.
5 • The left atrium is relaxed and receives oxygenated blood from the pulmonary vein.
  • When full the pressure forces open the bicuspid valve between the atrium and ventricle.
  • Relaxation of the left ventricle draws blood from the left atrium.
  • The left atrium contracts, pushing the remaining blood into the right ventricle through the valve.
  • With the left atrium relaxed and with the bicuspid valve closed the left ventricle contracts.
  • The strong muscular walls exert a strong pressure and push blood away from the heart through the semilunar valves through the pulmonary arteries and the aorta.

Exam practice

1 cardiac; myogenic; sinoatrial node; atrio-ventricular node; bundle of His; Purkinje/Purkyne fibres; contract; pulmonary artery.

Transport in plants

Checkpoints

1 In the stem the vascular tissue is found on the periphery; in the root it is found in the centre.
  In the stem xylem and phloem are on the same radius; in the root they are on alternate radii.
2 In the stem the peripheral position counteracts a bending strain; in the leaf the network of veins resists the tearing effect of wind; in the root the central position resists pulling by wind.
3 The sub-stomatal chamber has a high water potential as the walls of the spongy mesophyll cells are saturated with water. The water evaporates from the walls and moves down a gradient of potential from the plant to the atmosphere.
4 If a plant loses more water through transpiration than it can take up into its roots.
5 The water potential gradient is reduced as the stomata are enclosed within a 'microclimate' with epidermal hairs helping to maintain high humidity.

Exam practice

1 A reference should be made to the following methods of reducing evaporation: rolled shape giving reduced surface area; thick waxy cuticle impervious to water; stomata sunken in pits with chamber above stoma becoming saturated with water.

Transport of water through the plant

Checkpoints

1 The deposition of lignin between cellulose microfibrils in the cell wall.
2 Anchorage.
3 Passive.
4 Sap exudes due to root pressure.
5 Oxygen and glucose providing energy by respiration in the mitochondrion.
6 Mineral uptake reduced or stopped as it is energy requiring.

Exam practice

1 (a) (i) A, vacuole; B, cytoplasm; C, Casparian strip; D, xylem.
   (ii) Increase surface area for uptake of minerals and water; anchorage.
   (iii) It prevents water moving along the apoplast pathway and so means everything must go through the symplast pathway.
   (b) There is a mass flow of water pulled by evaporation from the leaves and water molecules are held together by cohesive forces.

Translocation in plants

Checkpoints

1 Sucrose and amino acids.
2 Sieve tubes in a leaf are a 'source' region for pressure flow. Phloem in leaves contains a high concentration of sucrose and therefore will have a low (more negative) water potential. Water flows into the phloem from the xylem and a high hydrostatic pressure is created.

Roots are a 'sink' region. In the cells of a root sugars are turned to starch. There will be a relatively high (less negative) water potential. Water flows into neighbouring cells by osmosis and a low hydrostatic pressure develops.
3 Growing points, food storage organs.

Exam practice

1 (a) The diagrams needed are of phloem tissue, a sieve tube and a companion cell; a diagram showing the relationship of the vascular tissue with transfer cells could be helpful also. Give a written description of these cells and their structures as well.
   (b) One of the following experiments should be described: removal of a ring of bark, using radioactive tracers or using aphid mouth parts.
2 (a) (i) Xylem: tracheids or vessels transport materials upwards from roots to leaves by transpiration pull (stream) or cohesion–tension or root pressure. Phloem: sieve tubes (elements) transport materials up and down by mass flow or cytoplasmic streaming or transcellular strands or electro-osmosis.
   (ii) Allow aphid stylet to penetrate phloem, remove body, fluid oozes out or supply plant with radioactive-labelled CO_2 and follow sucrose using autoradiography or describe a ringing experiment.
   (b) Sucrose: synthesized in the leaves by photosynthesis; actively loaded; transported to sinks in the phloem which may be above or below (taken to other parts of the plant).
   Magnesium: imported from the soil to growing points or leaves in the xylem where some is used to manufacture chlorophyll; active transfer of magnesium from xylem to phloem for redistribution.

Revision checklist
Exchange and transport

By the end of this chapter you should be able to:

1	Distinguish between passive and active transport.	Confident	Not confident **Revise** page 39
2	Explain the meaning of osmosis in terms of water potential.	Confident	Not confident **Revise** page 38
3	Describe the opening and closing mechanism of stomatal action.	Confident	Not confident **Revise** page 43
4	Explain in terms of surface area to volume ratio why multicellular organisms need specialised exchange surfaces.	Confident	Not confident **Revise** pages 44
5	Analyse and interpret graphical data from a spirometer.	Confident	Not confident **Revise** page 45
6	Describe the role of haemoglobin in oxygen and carbon dioxide carriage.	Confident	Not confident **Revise** page 46
7	Illustrate the Bohr effect.	Confident	Not confident **Revise** page 47
8	Distinguish between: • Open and closed blood circulatory systems • Single and double blood circulatory systems	Confident	Not confident **Revise** page 48
9	Describe the cardiac cycle with reference to the action of valves in the heart.	Confident	Not confident **Revise** pages 50
10	Analyse graphical data showing pressure changes in the heart and blood vessels.	Confident	Not confident **Revise** page 50, 51
11	Describe the coordination of heart action with reference to SAN and AVN.	Confident	Not confident **Revise** page 51
12	Define the term xerophyte and explain the adaptations of these plants to reduce water loss.	Confident	Not confident **Revise** page 53
13	Describe the pathway by which water is transported through the plant.	Confident	Not confident **Revise** page 54
14	Outline the experimental evidence that supports the theory that translocation of solutes takes place in phloem tissue.	Confident	Not confident **Revise** page 56
15	Explain translocation as an energy requiring process transporting sucrose between sources and sinks.	Confident	Not confident **Revise** pages 56, 57

Energy and the environment

The study of the flow of energy through the ecosystem is known as ecological energetics. The source of energy in an ecosystem energy is light, and this is converted to chemical energy by the process of photosynthesis. The energy then passes along the food chain, at each stage of which some of the energy is lost as heat. Unlike energy, minerals such as carbon and nitrogen can be continuously recycled. Applied ecology requires information about the size of plant and animal populations so that humans can regulate and manage populations of organisms for their own use. The increasing pressures placed upon the natural environment have created problems and humans have come to understand the need for more careful conservation of resources.

Exam themes

- What is meant by an ecosystem?
- The efficiency of energy transfer between trophic levels
- The cycling of nitrogen within an ecosystem
- Global warming
- Factors affecting population growth
- The advantages and disadvantages of organic farming
- The implications of intensive food production in terms of the effect of farm waste, land reclamation and hedgerow destruction
- The human exploitation of resources and the need for conservation
- Deforestation
- The impact of agriculture on the environment
- Problems associated with the overuse of fertilizers

Topic checklist

	Edexcel		AQA		OCR		WJEC	
	AS	A2	AS	A2	AS	A2	AS	A2
Autotrophic and heterotrophic nutrition		●		●		●	○	
Energy flow in the ecosystem 1		●		●		●		●
Energy flow in the ecosystem 2		●		●		●		●
Food chains and food webs		●		●		●	○	
Recycling of nutrients 1		●		●	○			●
Recycling of nutrients 2		●		●		●		●
Population growth and control		●		●		●		●
Community and succession		●		●		●		●
Biodiversity and agriculture	○	●	○	●	○	●		●

Autotrophic and heterotrophic nutrition

Most autotrophic organisms use the simple organic materials, carbon dioxide and water, to manufacture energy-containing complex organic compounds, whereas heterotrophic organisms consume complex organic food material.

Autotrophic nutrition

There are two types of autotrophic nutrition.

Photosynthesis

This is the process by which green plants, algae and certain types of bacteria build up complex organic molecules from carbon dioxide, water and mineral ions. The source of energy for this process comes from sunlight which is absorbed by chlorophyll and related pigments.

Chemosynthesis

This is the process by which a few bacteria can perform similar synthesis of organic compounds using energy derived from special methods of respiration.

Heterotrophic nutrition

Heterotrophic organisms consume complex organic food material. There are a number of different forms of heterotrophic nutrition.

Holozoic

This involves taking in complex organic molecules, breaking them down by digestion, absorption into the body tissues from the digestive system and, finally, utilization of the absorbed products of digestion in the body cells. Animals that feed solely on plant material are termed **herbivores**, those that feed on other animals are **carnivores**, and the **omnivores** have a mixed diet.

Saprobiontic

This is also known as saprotrophic or saprophytic. It involves the absorption of complex organic food from the bodies of decaying organisms. Some bacteria and fungi feed in this way. They secrete enzymes on the food substrate and then absorb the products of this extracellular digestion. The activities of these organisms are important in the decomposition of leaf litter and the recycling of valuable nutrients.

Parasitic

A parasite is an organism that lives in or on another living organism, the host. The parasite derives all its nutrition from the host, whereas the host does not gain any benefit from the association and is often harmed to some degree. Some parasites live on the outside of the host, the **ectoparasites**, e.g. leech. Others live entirely within the body of the host and are termed **endoparasites**, e.g. Plasmodium, the malarial parasite.

Links

Photosynthesis is looked at in detail on page 92.

Checkpoint 1

List two differences between autotrophic and heterotrophic nutrition.

Links

Food chains are dealt with on pages 70 and 71.

Checkpoint 2

State one harmful effect of saprobionts.

Examiner's secrets

Examiners are impressed when candidates name specific examples.

Checkpoint 3

Why do most parasites produce very large numbers of eggs or spores?

Parasites are considered to be very highly specialized organisms and show considerable adaptations to their mode of life.

Mutualistic

This is also known as symbiosis. It also involves a close association between members of two species, but in this case both derive some benefit from the relationship, e.g. the digestion of cellulose by microorganisms in the gut of a herbivore.

Checkpoint 4

Give another example of mutualism.

Checkpoint 5

Which of the following types of organisms can live independently of their food supply: holozoic, saprobiontic, parasitic?

Exam practice   answers: page 82

1 The terms 'autotrophic', 'heterotrophic' and 'parasitic' are used to describe the ways in which different organisms obtain their nutrients. The table shows a list of organisms. Complete the table by ticking the box or boxes which apply to each organism. It may be necessary to tick more than one box to fully describe the way in which nutrients are obtained.

Organism	Autotroph	Heterotroph	Saprophyte	Parasite
Apple tree				
HIV				
Dog				
Nitrogen-fixing bacteria				
Penicillium				
Tapeworm				

2 Using the diagram below explain why
   (a) (i)   spore cases are found on long 'stalks'
       (ii) the 'bodies' (mycelium) of the fungus are always found in the surface layer of the substrate and not deep inside the tissues
       (iii) the mycelium covers a large area
   (b) What is meant by extracellular digestion?

Examiner's secrets

Don't tick too many boxes in the hope that you'll have covered the right ones. Examiners will penalize you for this.

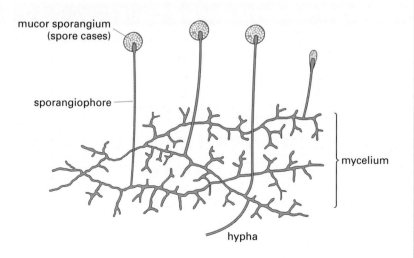

mucor sporangium (spore cases)

sporangiophore

mycelium

hypha

The jargon

A *mycelium* is a collection of hyphae.

Energy flow in the ecosystem 1

An ecosystem is a community of organisms that, together with their physical environment, form a self-perpetuating ecological unit. The study of the flow of energy through the ecosystem is known as ecological energetics.

Checkpoint 1

Which terms describe the following features of a seashore: all limpets on the seashore; temperature variations in a rock pool?

Components of an ecosystem

→ A *population* is a group of organisms of a single species occupying a particular area. In this area there will also be other populations, forming a community.
→ A *habitat* is the particular area occupied by a population. It has biotic and abiotic features that separate it from other habitats. The biotic features are the sum total of the organisms within the habitat and their interactions. Abiotic features include different types:
  → **edaphic** features relate to the soil and include all its physical and chemical characteristics;
  → **climatic** features include light, temperature, moisture, salinity and, particularly, the stability or variability of these.
→ *Microhabitats* are small localities within a habitat, each with its own particular conditions.
→ *Ecological niche* is the place of each species in an ecosytem. This is not only the physical space that it occupies, but the role which it carries out within the community and its interrelationships with other species as well. In the long term, two species cannot occupy the same niche in a specific habitat.

Checkpoint 2

What is the main difference between a niche and a habitat?

Energy source

The ultimate source of energy for ecosystems is the sun, from which energy is released in the form of electromagnetic waves. Only a small part of the total amount of energy reaching the Earth's atmosphere enters ecosystems. Also, the quantity absorbed by plants varies considerably at different latitudes. Of the energy entering a plant only 1–5% is utilized by the plant; the rest is lost, partly by reflection and partly by the evaporation of water.

Checkpoint 3

What is the ultimate fate of most energy on the Earth?

Energy flow through producers

The energy flowing from one organism to another in the food chain is not recycled but eventually lost as heat and so it must constantly be replaced by sunlight reaching the Earth. If it is assumed that 100 units of energy per unit time reach the leaves of a plant, the diagram at the top of the opposite page shows what happens to the energy.

Photosynthetic efficiency indicates the ability of a plant to trap light energy.

$$= \frac{\text{Quantity of energy incorporated into product}}{\text{Quantity of energy falling on the plant}}$$

This is normally less than 10% and depends on external factors such as light intensity and temperature.

fate of solar energy reaching the leaf of a crop plant (numbers are percentages of the total solar energy falling on the leaf)

lost in photosynthesis and evaporation   transmitted

Watch out!

Don't confuse NPP with the total biomass of autotrophs present at a given time, called the standing crop biomass. Primary productivity is the *rate* at which the organism synthesizes *new* biomass.

Checkpoint 4

If the NPP of grass = 21 135 kJ m^{-2} year^{-1} and a bullock eats 3 056 kJ m^{-2} year^{-1} what happens to the rest of the energy?

Gross primary productivity (GPP) is the rate at which products, such as glucose, are formed. A substantial amount of gross production is used up in respiration by the plant.

GPP – respiration = net production

That which is left over after respiration is called **net primary productivity** (NPP). This represents the food available to primary consumers. In crop plants this represents the yield which may be harvested. **Secondary productivity** is the rate at which consumers accumulate energy in the form of cells or tissues.

The jargon

Biological productivity is the rate at which biomass accumulates in an ecosystem.
Biomass is the dry weight of organic matter comprising a group of organisms in a particular habitat.

Exam practice   answers: page 82

1 The diagram below shows the quantity of energy flowing through a food chain in a terrestrial ecosystem. The figures given are kJ m^{-2} year^{-1}.

incident sunlight 3 × 10^6

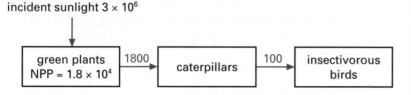

(a) Calculate the percentage of the incident energy which becomes available as the net primary production (NPP) of green plants. Show your working.

(b) Give two reasons why not all the energy of the incident sunlight is incorporated into the biomass of green plants.

(c) Using the information shown in the diagram, explain why the biomass of insectivorous birds is usually very much less than the biomass of caterpillars.

Examiner's secrets

Marks are allocated for showing the working in calculations. Don't do everything on the calculator giving only the answer.

Energy flow in the ecosystem 2

Transfer of energy from producers to primary consumers also involves considerable wastage of energy. Only about 10% of the energy available is transferred from one trophic level to the next. Carnivores are more efficient at energy conversion than are herbivores.

Energy flow through consumers

Consumers have a conversion efficiency of about 10%, that is, for every 100 grams of plant material taken in or ingested, only about 10 grams is incorporated into herbivore biomass. This means that only part of the NPP of the ecosystem is transferred to the primary consumers. Herbivores are not able to eat all the vegetation available to them. Consider a cow feeding on grass in a field.

→ Some of the plant material is not eaten by the cow. Cattle grazing a field will eat the grasses and edible weeds but do not eat the roots and often leave the woody parts of the plants.
→ Cows feed on plant material which contains cellulose which they are unable to digest. This passes out of the body as faeces containing a high proportion of undigested matter. (In terms of the ecosystem this energy is not wasted as it is available to decomposers.)
→ Some of the food material in the field is being eaten by other herbivores, such as rabbits.

Much of the energy in the food consumed by a cow is lost mainly by two processes:

→ Respiration – about 30% lost as heat
→ Excretion – about 60% lost in urine and faeces

Farmers increase the productivity of their livestock by keeping the animals in barns during the winter months because their movement is restricted and they lose less body heat.

Herbivores have a lower secondary productivity than carnivores. That is, carnivores are more efficient at energy conversion than herbivores. They have a much higher secondary productivity. This is because their protein-rich diet is more readily and efficiently digested. Only about 20% of the energy intake is lost in the faeces and urine of carnivores compared with a loss of about 60% in herbivores. Carnivores absorb almost twice as much energy per unit mass of food compared with herbivores.

The loss of energy at each trophic level gives ecological pyramids their characteristic shape.

Checkpoint 1
When food is in short supply explain why more vegetarians can be fed than those who rely on a meat-rich diet.

Checkpoint 2
Why do carnivores have a much higher secondary productivity than herbivores?

Checkpoint 3
Why do farmers keep their animals in barns during the winter months?

Checkpoint 4
Why does a cow seem to eat all day long, whereas a lion eats every few days?

Pyramids of energy

The number of organisms, their biomass or the amount of energy contained in each trophic level can be represented in diagrams with a bar for each level. These are known as pyramids. They provide a quantitative account of the feeding relationships in a community. The most accurate way to represent the feeding relationships in a community is to use a **pyramid of energy**.

A pyramid of energy shows the quantity of energy transferred from one trophic level to the next, per unit area or volume, per unit time. This represents the total **energy** requirement of each successive trophic level in a food chain. As material passes up through the food chain, energy is lost in respiration as heat, and in excretion, so the size of the bars decreases sharply. Since only some of the energy is passed on from one level to the next, energy pyramids are never inverted as in biomass pyramids.

The use of a set period of time means that an energy pyramid overcomes the problems which arise when ecosystems are compared simply by counting or measuring the standing crop of organisms.

Pyramids of energy enable ease of comparison of the efficiency of energy transfer to be made from one trophic level to the next between different communities. However, obtaining the data can be complex and difficult.

Exam practice   answers: page 82

1 The diagram shows the energy input, output, and the amount of energy present at each trophic level of an aquatic ecosystem. The figures represent the amount of energy per square metre over a year.

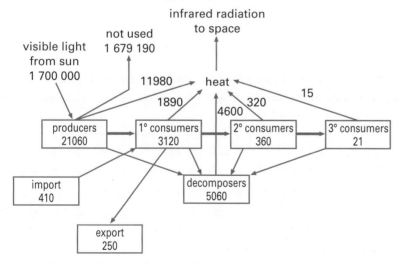

(a) (i) Calculate the percentage energy from the sun converted by the producers during photosynthesis.

   (ii) Give 3 reasons why most of the energy from the sun (1 679 190) is 'not used'.

(b) (i) How many trophic levels are shown in this ecosystem?

   (ii) Explain why a further trophic level is not possible in this ecosystem.

(c) Energy is 'lost' from the producers and consumers in the form of heat. Give 2 other ways in which some of the energy is 'lost'.

(d) This ecosystem is a small lake. Suggest how energy may be exported.

Checkpoint 5

Define 'pyramid of energy'.

The jargon

Biomass is the mass of living organisms present at a given time.

Grade booster

Be able to explain the difference between warm-blooded and cold-blooded animals in terms of energy requirement.

Checkpoint 6

How is the energy of an ecosystem measured?

The jargon

Standing crop describes the amount of living material present at a given instant in time.

Food chains and food webs

Energy is passed from one feeding level (or trophic level) to another through the ecosystem. Energy is passed along a hierarchy of trophic levels with primary producers at the bottom and consumers at the top. This is referred to as a food chain. However, in practice, organisms feed off more than one type of food. Therefore, feeding relationships are better described as a complex series of interconnections known as a food web.

Checkpoint 1

What is the ultimate source of energy of a food web?

Watch out!

Candidates often perform poorly in ecology because they do not learn the definitions of the terms used.

Food chains and food webs

→ Green plants which manufacture sugars from simple raw materials using solar energy are called **primary producers**.
→ Animals feeding on these plants are called herbivores or **primary consumers**.
→ Animals feeding on these animals are called carnivores or **secondary** and **tertiary consumers**.

Each of these groups forms a feeding or trophic level, with energy passing from each level to a higher one as material is eaten. At each level energy is lost through respiration and in waste products, so the amount of energy is reduced.

The sequence from plant to herbivore to carnivore is a food chain and is the route by which energy passes between trophic levels. It is the loss of energy at each level which limits the length of a **food chain** so the number of links in a chain is normally limited to four or five.

On the death of producers and consumers, some energy remains locked up in the organic compounds of which they are made. **Decomposers** feed as saprobionts and contribute to the recycling of nutrients (see diagram below).

Checkpoint 2

What is the difference between a decomposer and a detritivore?

Links

Recycling of nutrients is looked at on pages 72–75.

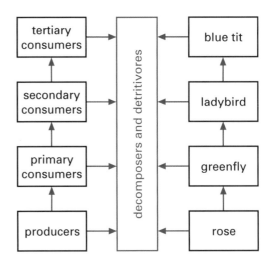

Checkpoint 3

What is the main difference between a food chain and a food web?

Watch out!

A single species may form part of many different chains, not always occupying the same trophic level in each chain.

Single food chains rarely exist and a **food web** is a more realistic representation of feeding relationships. This is because most primary consumers feed on more than one kind of autotroph and most secondary consumers rely on more than one type of prey. The more varied the organisms in an environment, the more complex the food web.

Trophic efficiency

This is the percentage of energy at one trophic level which is incorporated into the next trophic level. The rate at which energy passes into the animals at each trophic level is about 10% of that entering the previous level. This is called the gross ecological efficiency. This value differs from one ecosystem to another, with some of the highest values, around 40%, occurring in oceanic food chains. Some of the lowest values, around 1%, are found in ecosystems where most of the animals are birds or mammals.

Checkpoint 4

What happens to the other 90% of the energy?

Checkpoint 5

Why is the trophic efficiency of small mammals generally less than 1%?

Exam practice   answers: page 82

1   The diagram below shows part of a food web in a freshwater pond. Name different organisms from the web which provide examples of:

(a) an omnivore

(b) a producer

(c) a secondary consumer

(d) a tertiary consumer

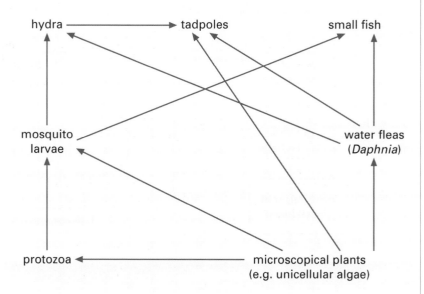

2   (a) Plants in a grassland ecosystem assimilate about 1% of the sun's radiant energy that falls on them. Approximately 80% of the plants' assimilated energy is available to the next trophic level.

(i) Give two reasons why plants assimilate so little of the sun's energy.

(ii) What happens to the energy assimilated by plants that is not available to the next trophic level?

(b) Explain why the number of links in food chains does not often exceed five.

Action point

For a habitat other than a freshwater pond, construct a food web by interconnecting all the organisms using arrows. Don't forget that the arrow heads indicate the direction of energy flow.

Watch out!

Be aware that the same organism can occupy different trophic levels in a food web. For example, in the diagram opposite a tadpole is a primary consumer when it feeds on algae but a secondary consumer when it feeds on water fleas.

Recycling of nutrients 1

Checkpoint 1

How has the level of CO_2 in the atmosphere been maintained?

Action point

The combustion of fossil fuels and the burning of enormous quantities of wood removed by deforestation has resulted in an increase in the concentration of carbon dioxide in the atmosphere. Plan an essay that considers the effects of these human activities.

All living organisms require carbon. Organic molecules such as carbohydrates proteins and fats contain carbon. The level of CO_2 in the atmosphere has been constant for millions of years. In recent times the level of CO_2 has increased as a result of human activity. The possible consequence of this in terms of the 'greenhouse effect' is global warming.

Carbon cycle

Carbon dioxide is added to the air by the respiration of animals, plants and microorganisms and by the combustion of fossil fuels. Photosynthesis takes place on so great a scale that it re-uses on a daily basis almost as much carbon dioxide as is released into the atmosphere. In past times large quantities of dead organisms accumulated in anaerobic conditions and so were prevented from decaying. In time they formed coal, oil and other fossil fuels. The burning of these fuels returns more CO_2 to the atmosphere and has resulted in a rise in CO_2 in the atmosphere, particularly over the last 50 years.

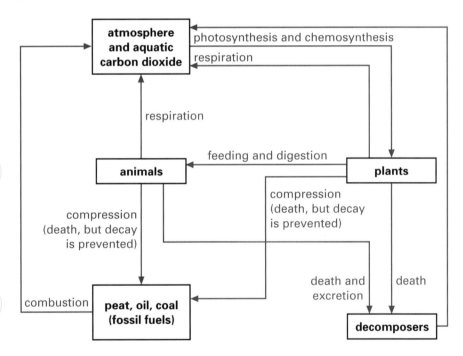

Don't forget

Nitrogen is found in all amino acids which make up proteins. Nitrogen is available to plants only in the form of ammonium ions (NH_4^+) and nitrate (NO_3^-) ions which are taken up by the roots.

Links

Deforestation on page 80–81.

Watch out!

Plants respire all the time just as animals do! During high light intensity the rate of photosynthesis is greater than the rate of respiration so the overall gas released is oxygen.

Humans and the carbon cycle

There are two main reasons for the rise in CO_2:

→ **burning of fossil fuel** - accounts for about 70% of the increase in CO_2.

→ **deforestation** – accounts for about 30% of the increased CO_2 levels, as about half the world's forests having been removed over the last 30 years.

Gases such as CFC's, nitrous oxides, methane and CO_2 are called **greenhouse gases** as they form a layer in the atmosphere acting like glass in a greenhouse. CO_2 is particularly important in the greenhouse effect because it represents about 49% of all greenhouse gases.

Greenhouse gases allow high-energy solar radiation to pass through to the Earth's surface. Much of this energy is 'bounced back' towards space as heat but some of the energy is absorbed and trapped by the greenhouse gases. The greenhouse effect is a natural process without which the average temperature on Earth would be too low to sustain life.

The increase in greenhouse gases is thought to be the cause of '**global warming**'.

Global warming

Global warming refers to an average increase in the Earth's temperature, which in turn causes changes in climate. A warmer Earth may lead to changes in rainfall patterns, a rise in sea level and a wide range of impacts on plants, wildlife, and humans.

Possible consequence of global warming may be:

→ Some melting of polar icecaps resulting in flooding in coastal areas
→ Increased frequency of droughts, hurricanes and cyclones, and also forest fires
→ In tropical areas of the world decreased availability of water might lead to the formation of deserts
→ Massive reductions in the grain crops of North America and Central Asia. This would have serious economic and political consequences.

> **Checkpoint 2**
>
> What climatic change has taken place due to the removal of trees in tropical rain forests??

Carbon footprint

The term 'carbon footprint' owes its origins to the idea that a footprint is what has been left behind as a result of an individual's activities. There are two possible ways of 'tracing' a carbon footprint:

→ the total amount of CO_2 and other greenhouse gases emitted over the full life cycle of a **product or service**. Normally a carbon footprint is usually expressed as a CO_2 equivalent (usually in kilogrammes or tonnes)
→ the total amount of CO_2 attributable to the actions of an **individual** (mainly through their energy use) over a period of one year

> **Checkpoint 3**
>
> What is meant by the carbon footprint?

Individual carbon footprints can be either:

→ direct emissions (typically from energy used in the home and in transport)
→ indirect emissions, including CO_2 emissions as a result of goods and services consumed

> **Exam practice**   answers: page 83
>
> 1 The diagram shows the carbon cycle.
>
>
> (a) Name the processes labelled 1–4.
> (b) In what form are carbon atoms most commonly found in the atmosphere?

> **Grade booster**
>
> Evaluate the evidence that global warming is taking place. What actions are governments taking?

Recycling of nutrients 2

Living organisms need nitrogen to make amino acids, proteins and nucleic acids. The nitrogen cycle is the flow of organic and inorganic nitrogen within an ecosystem where there is an interchange between nitrogenous compounds and atmospheric nitrogen. Microorganisms play a key role in the process of decay.

Nitrogen cycle

Plants and animals are unable to use nitrogen gas. Instead plants take in nitrates in solution through their roots. The organic nitrogen compounds produced by plants are transferred through the food chain when consumers eat plants. When plants and animals die the minerals locked in their bodies, together with the excretory products of animals, must be decomposed in order to release the minerals back into the soil.

The main processes involved are as follows:

→ **Putrefaction** – bacteria and fungi decompose dead plants and animals, faeces and urine into ammonium ions. This process is also called ammonification.

→ **Nitrification** – the ammonia formed in putrefaction is converted by nitrification via nitrites to nitrates. Various bacteria are involved. For example, ammonia is converted to nitrite by *Nitrosomonas* and nitrite to nitrate by *Nitrobacter*. These bacteria require aerobic conditions.

→ **Nitrogen fixation** – atmospheric nitrogen can be converted directly into nitrogen compounds by nitrogen-fixing bacteria. Free living nitrogen-fixing bacteria include *Azotobacter*. These account for most of the nitrogen fixation. There are also symbiotic nitrogen-fixing bacteria, *Rhizobium*, found in the root nodules of legumes (peas, beans and clover).

→ **Denitrification** – nitrogen is lost from ecosystems by denitrification. This is a particular problem in waterlogged soils with anaerobic conditions where anaerobic bacteria, such as *Pseudomonas*, can reduce nitrates and ammonium ions back to nitrogen.

Human activities can improve the circulation of nitrogen.

→ By fixing atmospheric nitrogen artificially using chemical processes that convert it to fertilizers
→ Large amount of animal waste from stock rearing is used as manure
→ Sewage disposal boosts organic nitrogen supplies
→ Microorganisms can be used for making compost and silage
→ Farming practices such as:
   → planting fields of clover to encourage nitrogen fixation
   → draining land and reducing anaerobic conditions
   → ploughing fields in order to improve aeration of the soil

---

Checkpoint 1

Is the conversion of ammonium to nitrite an oxidation or a reduction process?

Checkpoint 2

Why do nitrifying bacteria thrive in waterlogged soils?

Checkpoint 3

What is the link between deforestation and the nitrogen cycle?

Watch out!

You should know the specific names of the microbes: *Nitrosomonas*, *Nitrobacter* and *Rhizobium*.

Grade booster

Candidates often find the nitrogen cycle confusing, so make sure you are clear about it.

Don't forget

Decomposition accounts for most of the conversion of organic material from all trophic levels into inorganic compounds that can be recycled.

Problems caused by nitrate-containing fertilizers

As the human population has increased, people's expectations to have plentiful, high-quality food have risen. Agricultural practices have had to change in an attempt to meet the demands. In developed countries agriculture has become more intensive, providing high crop yields from relatively small areas of land. However, this increase in food production has had a deleterious effect on the environment. The increased use of nitrate-containing fertilizers has had some harmful effects on both aquatic and terrestrial ecosystems.

→ Problems caused by excess nitrate in soils:
  On agricultural land the increased use of fertilizer has reduced species diversity on grassland. Fertilizers increase the growth of grasses and plants such as nettles, which shade out smaller plants.

→ Problems caused by nitrates leaching into rivers:
  The leaching of salts from the surrounding land is a slow, natural process. However, sewage and fertilizers are an additional source of these salts.
  → Nitrate is highly soluble and is readily **leached** from soil and washed into rivers from surrounding land.
  → The first effect may be an **algal bloom**, where the waters become densely populated with species of algae, their growth being enhanced by abundant nutrients.
  → At this stage the water may become green and light is unable to penetrate to any depth. The plants in the deeper regions of the lake are unable to photosynthesise and die. There is a general decrease in animal species diversity as they rely on the plants for food and shelter.
  → The short-lived algae soon die and are decomposed by saprobiontic bacteria creating a biochemical oxygen demand (BOD). The water in all but the very upper layers becomes deoxygenated, so that fish and other oxygen-requiring species die.
  → In the final stages of the process of **eutrophication**, anaerobic bacteria in the water may reduce nitrate to nitrite (both nitrate and nitrite are toxic compounds). In view of their toxicity, the EU has set a limit of 11.3 parts per million (ppm) total nitrogen in drinking water. This figure has been exceeded in parts of the UK.

Checkpoint 4

How has the use of fertilizers decreased the biodiversity of grassland?

Checkpoint 5

What is an algal bloom?

Checkpoint 6

Define 'eutrophication'?

Don't forget

Nitrogen is found in all amino acids which make up proteins. Nitrogen is available to plants only in the form of ammonium ions (NH_4^+) and nitrate (NO_3^-) ions which are taken up by the roots.

The jargon

Biochemical oxygen demand (BOD) is the rate at which oxygen is used up in a sample of water. A high BOD indicates a rapid rate of aerobic respiration by bacteria.

Grade booster

The overuse of fertilizers has had a major effect on biodiversity. Research this issue.

Exam practice   answers: page 83

1 (a) Give a full account of the ways in which nitrogen is exchanged between organisms and their environment.

  (b) Discuss the various ways in which human activities influence the nitrogen cycle.

Population growth and control

Organisms live as part of populations and communities. You should understand how factors can limit the growth of a population and how the balance between birth rate and death rate is maintained. It is possible to study the development of a community over time.

Population growth

A typical S-shaped curve of population growth traces the increase in number of a species moving into a new geographical area (see graph below).

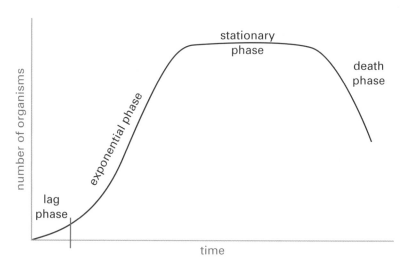

Checkpoint 1

Explain the reason for the lag phase when yeast is introduced into a solution of glucose.

→ The *lag phase*: as only a few individuals are present initially, the rate of growth is very slow.
→ The *exponential phase*: as numbers increase, providing there is no factor limiting growth, more individuals become available for reproduction.
→ *Stationary phase*: certain factors limit the population growth and birth rate equals death rate.
→ *Carrying capacity*: this is the limit to the number of individuals that an area can support.

Action point

Draw a line on the graph above to show the carrying capacity of the population.

The factors that limit the growth of a particular population are collectively called the **environmental resistance**. Such factors include available food, predation, overcrowding, disease and accumulation of toxic waste.

Density-dependence

Some factors are density-dependent, i.e. their effect increases as the density of the population increases, e.g. accumulation of toxic waste, disease, parasitism and sometimes food supply.

Other factors are density-independent, i.e. their effect does not depend on the population density – the effect is the same regardless of the size of the population. It is usually due to a sudden or violent change in an abiotic factor, e.g. freezing, flood or fires.

Regulation of population size

The size of a population is regulated by the balance between the birth rate and the death rate. However, there are other influences on the size of a population.

→ *Immigration* occurs when individuals join a population from neighbouring populations.
→ *Emigration* occurs when individuals depart from a population.
→ *Competition* occurs when individuals of a species in a population are continually competing with each other. Plants compete for space, light and mineral ions; animals compete for food, shelter and a mate.

There are two forms of competition:

→ *intraspecific competition* occurs between individuals of the same species
→ *interspecific competition* occurs between individuals of different species

Predation

A predator is an organism that feeds on living species. Predators are normally larger than their prey and tend to kill before they eat. The abundance of prey is a factor limiting the numbers of the predator. Within a food chain, a predator–prey relationship causes both populations to oscillate and these oscillations are regulated by the process known as negative feedback.

Checkpoint 2

How can two species living in the same habitat avoid competition?

Checkpoint 3

What type of competition exists between a cormorant and a seagull?

Speed learning

It's very easy to confuse *intraspecific* and *interspecific* competition. Think of *international* sports matches **between different countries** – *interspecific competition* occurs **between** individuals of **different species**!

Exam practice   answers: page 83

1 The graph shows a typical growth curve for a population of organisms.

Checkpoint 4

What is the value of a negative feedback system in population control?

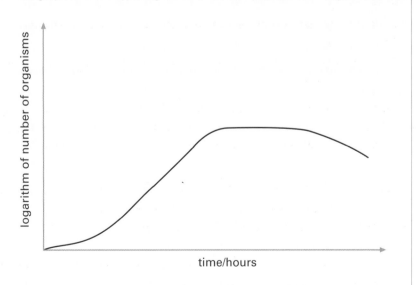

(a) Indicate and label clearly on the graph the 4 stages of a growth curve.
(b) List 3 factors which may limit the growth of a population, at least 1 of which must be a density-independent factor.
(c) Describe how each brings about the effect on the population.

Community and succession

The distribution of species does not necessarily remain the same over long periods of time. Ecosystems are dynamic and subject to change. Organisms and their environment interact: if one changes, so does the other. A change in the environment affects the organisms, and a change in the organisms affects the environment. Succession is the change in structure and composition of species over time.

Community and succession

An area of bare ground does not remain free of vegetation for long, as any gardener knows! Weeds are usually the first plants to grow, followed by grasses then taller plants. Provided the soil conditions are suitable, over a very long period of time the bare ground would eventually become woodland.

Checkpoint 1

What is the main difference between a primary and secondary succession?

→ **Primary succession** refers to the introduction of plants/animals into areas that have not previously supported a community, for example, bare rock, or the site of volcanic eruption.

Checkpoint 2

Give an example of a secondary succession.

→ **Secondary succession** refers to the reintroduction of organisms into a bare habitat previously occupied by plants and animals. If the original vegetation is removed, for example, by fire or tree felling, the area rapidly becomes recolonized by a succession of plants and animals.

In any area, over time, new organisms replace existing ones, that is, species diversity increases, until a stable state is reached. All successions usually involve changes in community structure and function until a community reaches a climax of succession known as the **climax community**, for example, a mature woodland. This is a stable community which undergoes no further change.

Checkpoint 3

What is the natural climax community in the UK?

The different stages in a succession are known as **seres**. Consider the colonization of bare rock.

The first organisms to colonise the bare rock are **algae and lichens**. These plants are called **pioneer species** and form a pioneer community. Lichens slowly erode the rock. This, together with the weathering of the rock and the accumulation of dead and decomposing organic material, leads to the formation of a primitive soil. Wind-blown spores allow **mosses** to appear and as the soil develops, grasses become established. As the soil builds up, deep-rooted **shrubs** appear. Over a very long time **trees**, such as oak, become established. This is then known as the climax community. This is a stable condition, dominated by long-lived plants.

The jargon

A *community* is defined as 'all the living organisms in a habitat'.

Checkpoint 4

What is a sere?

It should be noted that a community consists of animals as well as plants and that the animals have undergone a similar succession dictated by the plant types present at each stage.

In a secondary succession, seeds, spores and organs of vegetative reproduction may remain in the soil and dispersal of plants and migration of animals will assist in colonization of the habitat.

Human interference can affect a succession and may prevent the natural development of the climax community.

→ Grazing by sheep

→ Heather moorland management by controlled burning for grouse shooting

→ Farming of land

→ Deforestation and soil erosion

Grade booster

Research the succession of a specific community, such as sand dunes, and consider how human activity affects that community.

Exam practice   answers: page 83

1 Volcanic eruptions under the surface of the sea sometimes produce new islands such as Surtsey, 40 km off the coast of Iceland. In 1850 the island of Krakatoa, located 25 km off Java in Asia, suffered a violent volcanic eruption and all life on the island was destroyed. The table shows the number of flowering plant species that colonized the islands in the 12 years after the eruptions.

Number of years after eruption	Number of flowering plant species	
	Surtsey	Krakatoa
3	2	14
6	7	31
9	15	41
12	26	48

(a) What is meant by the term succession?

(b) Suggest 2 reasons for the faster rate of colonization on Krakatoa.

(c) The pioneer plants became established on the bare rock of the islands after the eruptions. Suggest the process by which these plants helped to make the conditions more suitable for other plants.

(d) Give 2 reasons why the increase in the number and variety of flowering plants allowed the diversity of insects to increase.

Biodiversity and agriculture

Conflicts can sometimes exist between farming and conservation. The main purpose of agriculture is to produce food for human consumption. Both the efficiency and the intensity of food production are being continually increased to meet the demands of the human population. To obtain the maximum possible yield from the land, changes in farming methods have taken place.

Checkpoint 1

State 2 advantages of large-scale monoculture.

Checkpoint 2

In terms of biodiversity what is the main disadvantage of large-scale monoculture?

Action point

Find out the changes that have taken place in farming methods during the last 50 years.

Grade booster

Research the Environment Act of 1995 and its implications for maintaining biodiversity.

Links

Eutrophication on page 75.

The jargon

Soil erosion is the removal of topsoil containing valuable nutrients.

Checkpoint 3

How does deforestation lead to soil erosion?

Changes in farming methods

These have a number of environmental implications:

→ To make **larger fields** many **hedgerows were removed** – larger machinery that was needed to prepare the soil and harvest crops could be used more efficiently.

→ Single crops were grown, for example, wheat and barley. This is called monoculture and is the simultaneous growth of large numbers of crop plants of similar age and type within a defined area. Monoculture is an 'artificial' situation because there will normally be a succession leading to a greater diversity of species. If the same crop is grown on the same plot year after year, yield progressively declines. This is due in part to mineral depletion but also because conditions become ideal for the crop plant's pests and parasites. Selective **herbicides** to prevent the growth of weeds, and **pesticides** to remove insect and other pests, were introduced.

→ Intensive cultivation necessitated an increase in the use of inorganic fertilizers which had unforeseen environmental effects, such as eutrophication of waterways.

In recent years the views of government, farmers and consumers have changed.

→ People are far more aware of the value of the countryside not only because it is a source of food but also because it provides a habitat for plants and animals as well as a place to visit for relaxation and enjoyment.

→ Schemes are in place to encourage farmers to manage their farms for biodiversity.

Deforestation

Forests cover about 34% of the world's land surface. However, about half the world's forests have been cut down by deforestation during the last 30 years. The sale of valuable timber, the freeing of land for alternative uses such as subsistence farming and cash crops, clearing land for roads, etc. has meant that the trees of forests and woodland are being cut down faster than they can be replanted or regenerated naturally.

Reasons for deforestation:

→ There is a world demand for timber as a building material.
→ Wood is used as a fuel.
→ Land is cleared for farming.

→ New roads are built to provide a transport infrastructure.
→ There is a demand for paper and packaging.

Grade booster

Research the reasons for and the extent of the destruction of tropical rain forests.

Consequences of deforestation:

→ **Climate change**

The rate at which carbon dioxide is removed from the atmosphere by the process of photosynthesis is being reduced by cutting down forests. On a global scale this is a massive reduction and contributes to global warming.

→ **Destruction of natural habitats**, leading to a reduction in biodiversity. It is estimated that at least 50% of the Earth's species live in the tropical rain forests, even though they occupy only about 10% of the Earth's land area. If natural habitats are destroyed this may lead to the loss of medicinal remedies – some tropical plants may become extinct before their clinical properties have been investigated.

→ **Soil erosion**
  → Digging and ploughing loosens the topsoil, assisting in the process of soil erosion.
  → The removal of vegetation affects regional climate mainly by reducing rainfall thus accelerating desertification.
  → Deforestation of the watershed causes lowland **flooding**. The removal of vegetation on the higher slopes of valleys results in heavy rain sweeping exposed soil to the flood plains below.

Exam practice   answers: page 83

1 The practice of clearing large tracts of forest (deforestation) to plant crops is now becoming increasingly widespread in the developing world.

   (a) State 2 reasons for deforestation.

   (b) Diagram I shows the result of rainfall on the natural vegetation on the steep side of a valley. Diagram II shows the same slope cleared of trees and planted with maize.

   Key
   Relatively heavy flow of water
   Relatively light flow of water
   Water runs off
   Water runs off
   Humus
   Diagram I
   Diagram II

      (i)   Using only the information in the diagrams suggest 2 different reasons why the force of water running down the slope is less in diagram I than in diagram II.

      (ii) Suggest 2 possible harmful consequences of the water runoff in diagram II to the people living in the valley.

   (c) (i)   Describe 1 possible long-term effect of deforestation on the forest-dwelling animals.

        (ii) Suggest 2 possible explanations of the effect you have described.

Answers
Energy and the environment

Autotrophic and heterotrophic nutrition

Checkpoints

1 Autotrophic: energy in; synthesis of organic compounds. Heterotrophic: energy out; breakdown of organic material.
2 Food spoilage.
3 To increase chance of finding another host.
4 Lichens, etc.
5 Holozoic.

Exam practice

1 Apple tree (A); HIV (H, P); dog (H); nitrogen-fixing bacteria (A); *Penicillium* (H, S); tapeworm (H, P).
2 (a) (i) So that spores are carried a greater distance by wind currents.
       (ii) They are aerobic and obtain oxygen on the surface.
       (iii) To increase surface area for release of enzymes.
   (b) Digestion occurs outside the cells; digested products are absorbed into the body.

Energy flow in the ecosystem 1

Checkpoints

1 Limpets – population; temperature – abiotic.
2 Niche: position occupied and lifestyle.
   Habitat: position occupied only.
3 Passed out into space.
4 Passed on to other herbivores and decomposers.

Exam practice

1 (a) (1.8×10^4 divided by 3×10^6) x 100 (for %) = 0.6.
   (b) Some of the energy is of the wrong wavelength; some is reflected.
   (c) Energy is lost at each trophic level so the biomass of animals in the higher levels is less (use the figures 100 and 1800 appropriately in your answer).

Energy flow in the ecosystem 2

Checkpoints

1 The amount of energy decreases by up to 90% at each step in the food chain so more energy is available at the lower levels.

2 A protein-rich diet is more readily and efficiently digested; no energy-consuming symbionts in their digestive tract; faeces contain less undigested matter since no cellulose in diet.
3 To restrict the movement of animals and reduce heat loss. This results in less energy loss in respiration.
4 A protein-rich diet is more readily and efficiently digested. Only about 20% of the energy intake of carnivores is lost in excretion.
5 The amount of energy transferred from one trophic level to the next per unit area or volume per unit time.
6 Using a calorimeter.

Exam practice

1 (a) (i) $\dfrac{21060 \times 100}{1\ 700\ 000} = 1.24\%$

       (ii) Wrong wavelength, reflected, absorbed by water, light passes through leaf and misses chloroplasts.
   (b) (i) 4.
       (ii) Energy at last trophic level is only 21 units which is insufficient to support another level.
   (c) Death/decay; excretion/urine/faeces.
   (d) Fishing.

Food chains and food webs

Checkpoints

1 Sunlight.
2 Decomposers are saprobiontic microorganisms; detritivores are larger animals, such as earthworms, which digest food internally rather than externally.
3 Some consumers feed on more than one kind of organism and a food web is therefore a more realistic representation of feeding relationships.
4 Lost in metabolism, respiration and excretion.
5 Small mammals have a large surface area to volume ratio and as a result they lose heat rapidly. An increased metabolic rate is needed to generate heat and maintain body temperature.

Exam practice

1 (a) Mosquito larva, tadpole.
   (b) Algae.
   (c) Fish, mosquito larva, hydra, tadpole.
   (d) Hydra, fish, tadpole.
2 (a) (i) Energy is reflected from the shiny cuticle of the leaf; some light passes through the leaf without being absorbed; only some of the energy is of the correct wavelength for photosynthesis.
       (ii) It is used by the plant itself for respiration; some is lost as heat.
   (b) Only about 10% of the energy at any trophic level is available to the next, the rest is lost in metabolism. Only a proportion of the energy is therefore passed on and there is insufficient energy available to support more than five levels.

Examiner's secrets

In question 1 do not use the same animal more than once in your answers.

Recycling of nutrients 1

Checkpoints

1 There is a balance between the CO_2 added to the air by the respiration of animals and microorganisms and the removal of CO_2 by plants using CO_2 in photosynthesis.
2 There has been a 30% increase in carbon dioxide, which contributes as a greenhouse gas to global warming.
3 The total amount of CO_2 attributed to the actions of an individual over a period of one year.

Exam practice

1 (a) 1 – photosynthesis; 2 – respiration; 3 – combustion; 4 – decay.
  (b) carbon dioxide.

Recycling of nutrients 2

Checkpoints

1 Oxidation.
2 They are anaerobic organisms and have no aerobic bacterial competitors.
3 Soil erosion and the leaching of nitrates.
4 High nitrates increase the growth of plants such as nettles, shading out smaller plants resulting in a reduction in wild flowers in meadows.
5 A large increase in the growth of algae due to high levels of minerals.
6 A process in which enrichment of water with nitrates leads to an increase in algae and bacteria and a reduction in oxygen content.

Exam practice

1 (a) Your answer should include a concise description of the nitrogen cycle including putrefaction, nitrification, nitrogen fixation and denitrification. A well-annotated diagram would help the description.
  (b) • Humans fix atmospheric nitrogen artificially by chemical processes which convert it into fertilizers.
     • Large amounts of animal waste from stock rearing is used as manure.
     • Sewage disposal boosts organic nitrogen supplies.
     • The deliberate exploitation of microorganisms, e.g. during composting, silage production, etc.
     • Overuse of fertilizers leads to eutrophication in waterways

Population growth and control

Checkpoints

1 The yeast is producing enzymes by protein synthesis.

2 They occupy different niches, e.g. different feeding niches means they have a different diet.
3 Interspecific competition.
4 It maintains the numbers of organisms in the food chain balanced at levels that the environment can support.

Exam practice

1 (a) Lag; log/growth/exponential; stationary; death.
  (b)

Factor	Effect
Lack of food/water	Insufficient resources
Predation	Large numbers eaten by predators
Parasitism/disease	Parasites reduce reproduction of population
Stress	Reduces reproductive capability
Toxic wastes produced	Poisons organisms
Climate change/ environmental catastrophe (DI)	Kills large numbers of population

Community and succession

Checkpoints

1 Primary succession involves the introduction of plants and animals into an area that has not previously supported a community, whereas secondary succession refers to the reintroduction of organisms into a bare habitat previously occupied.
2 An area destroyed by fire.
3 Oak woodland.
4 A stage in succession.

Exam practice

1 (a) The sequence of alteration of communities with time.
  (b) Less distance for seeds to travel to Krakatoa; greater diversity on neighbouring tropical islands, reference to warmer climate; secondary succession.
  (c) Death of plants; provide developing soil; breakdown of plant materials leads to nutrients; humus retains water.
  (d) Increase in number of niches, shelter, habitat; increase in variety of food.

Biodiversity and agriculture

Checkpoints

1 Ease of harvesting, easier preparation and maintenance, easier for combine harvesters to operate, greater yield so guaranteed market.
2 Reduction of gene pool; loss of hedgerows resulting in fewer natural predators.
3 Lack of roots binding soil together so rainfall washes soil away.

Exam practice

1 (a) Increasing population requires more food, fuel; commercial harvest of forest for wood and cash crops.

(b) (i) Canopy of leaves breaks force of water which now drips to the ground; trees have larger, more extensive roots and so absorb more water; humus absorbs moisture.

(ii) Soil erosion; flooding.

(c) (i) Decrease in population/extinction of some species.

(ii) Loss of food for leaf feeders; disturbance of food chains; loss of shelter; change of microclimate.

Revision checklist
Energy and the environment

By the end of this chapter you should be able to:

1	Define the terms producer, consumer, decomposer and trophic level.	Confident	Not confident **Revise** page 70
2	Describe how energy is transferred through the ecosystem.	Confident	Not confident **Revise** pages 66, 67
3	Discuss the efficiency of energy transfer between trophic levels.	Confident	Not confident **Revise** page 71
4	Explain how human activities have caused an increase in carbon dioxide levels in the atmosphere.	Confident	Not confident **Revise** pages 72, 73
5	Describe how microorganisms recycle nitrogen within the ecosystem.	Confident	Not confident **Revise** page 74
6	Discuss the problems caused by nitrate-containing fertilizers.	Confident	Not confident **Revise** page 75
7	Draw and label a typical population growth curve.	Confident	Not confident **Revise** page 76
8	Distinguish between density-dependent and density-independent factors.	Confident	Not confident **Revise** page 76
9	Using examples explain the terms intraspecific and interspecific competition.	Confident	Not confident **Revise** page 77
10	Describe one example of primary succession resulting in a climax community.	Confident	Not confident **Revise** page 78
11	With reference to deforestation discuss the economic, social and ethical reasons for the conservation of biological resources.	Confident	Not confident **Revise** pages 80, 81

Metabolism, respiration and photosynthesis

Sunlight energy is captured in chemical form by autotrophic organisms during the process of photosynthesis. The energy-containing molecules produced are useful for storage, but they cannot be directly used by cells to provide the required energy. The transfer of energy from carbohydrate molecules, fats and proteins to molecules like adenosine triphosphate, which can be utilized by cells, occurs during respiration. ATP is a reservoir of potential chemical energy and it works in metabolism by acting as a common intermediate, linking energy-requiring and energy-yielding reactions.

Exam themes

- The universal role of ATP as the energy currency of all living organisms
- The synthesis of ATP by the electron transport chain on the membranes of the mitochondria
- Comparison of energy efficiency of aerobic and anaerobic respiration
- The role of chlorophyll in trapping light energy
- The light-dependent and light-independent stages of photosynthesis
- Relate leaf structure to its role in photosynthesis
- The structure of the mitochondrion and chloroplast and their role in ATP production
- Respiratory quotient (RQ) and the relative energy values of respiratory substrates

Topic checklist

	Edexcel		AQA		OCR		WJEC	
	AS	A2	AS	A2	AS	A2	AS	A2
The role of ATP in metabolism			●		●		●	●
Aerobic respiration			●		●		●	●
Anaerobic respiration and energy budgets			●		●		●	●
Photosynthesis			●		●		●	●
Biochemistry of the light-dependent stage			●		●		●	●
Biochemistry of the light-independent stage			●		●		●	●

The role of ATP in metabolism

Adenosine triphosphate, ATP, is often described as the energy currency of cells. The energy released when respiratory substrates are broken down is used to build up molecules of ATP from adenosine diphosphate (ADP) and inorganic phosphate. When the molecules of ATP are hydrolyzed, energy is released for reactions where it is needed in the cells.

The structure of ATP

Don't forget

If all the energy from glucose were released at once, the temperature of the cells would increase considerably.
So when substrates are broken down in enzyme-catalyzed stages, small bursts of energy are released as ATP molecules.

ATP is a nucleotide consisting of an organic base, adenine, as carbon sugar, ribose, and a sequence of three phosphate groups linked together.

The importance of ATP

Checkpoint 1

Write the equation for the hydrolysis of ATP.

The hydrolysis of ATP to ADP is catalyzed by the enzyme **ATPase** and the removal of the terminal phosphate yields 30.7 kJ mol^{-1} of free energy. The addition of phosphate to ADP (phosphorylation) requires the same amount of energy. ATP is therefore a means of transferring free energy from energy-rich compounds to cellular reactions requiring it. There are two forms of phosphorylation:

→ *oxidative phosphorylation*, which occurs during cellular respiration in all aerobic cells
→ *photophosphorylation*, which occurs during photosynthesis in chlorophyll-containing cells

The uses of ATP

ATP is not stored so it has to be synthesized as required. The rate of synthesis keeps pace with demand. A metabolically active cell may require 2 million ATP molecules per second. ATP provides the necessary energy for:

Checkpoint 2

Give an example of a material synthesized in cells.

→ muscle contraction
→ synthesis of materials within cells
→ nerve transmission
→ active transport

The chemiosmotic theory of ATP synthesis

Links

The structure of the mitochondrion is given on page 87.

This theory proposes that the energy for ATP synthesis comes from the electrochemical gradient across a membrane. The same principle applies for ATP synthesis in chloroplasts and mitochondria.

Mitochondria have an inner phospholipid membrane, which is folded to form cristae that are lined with stalked bodies. Within the inner membrane there is a mechanism which actively transports H^+ ions (protons) from the matrix into the space between the inner and outer membranes of the mitochondrion. This creates an electrochemical gradient of protons across the inner membrane. The process occurs as follows (see diagram below).

→ The oxidation of glucose in the matrix is coupled to the reduction of a carrier.

→ The reduced carrier passes an electron to an associated pump in the membrane, becoming reoxidized and releasing protons into the matrix.

→ The charged pump then pumps protons into the space between the inner and outer membranes.

→ Electrons with a slightly reduced energy level are transferred by a carrier to the next pump (the series of carriers and pumps is known as the electron transport system).

→ A high concentration of protons builds up in the inter-membrane space.

→ The protons are only able to flow back into the matrix via channels formed by ATPase molecules in the stalked bodies.

→ This flow acts as an energy source to combine ADP and inorganic phosphate to synthesize ATP.

→ The electrons from the last pump are taken up by oxygen, which combines with protons to form water. This is catalyzed by the enzyme oxidase.

In chloroplasts the build-up of protons is inside the membrane, whereas in mitochondria the build-up of protons is outside the membrane.

Checkpoint 3

What is the purpose of the cristae?

The jargon

The *pumps* in this context are large enzyme molecules that traverse the membrane.

Checkpoint 4

Why does ATP synthesis by this method not occur in anaerobic respiration?

Checkpoint 5

What name is given to the overall process of ATP production by this method?

Action point

Rephrase this list of points as essay-style paragraphs.

Links

ATP synthesis is relevant to aerobic respiration and photosynthesis, described on pages 88, 89 and 90, 91 respectively.

Grade booster

Make sure you have a good understanding of the chemiosmotic theory.

Exam practice   answers: page 98

1 Describe the electron transport system and explain fully how the structure of the mitochondrion:

(a) enables the electron transport chain to function

(b) ensures the synthesis of ATP

Aerobic respiration

Respiration is a series of enzyme-catalyzed reactions that release energy from organic molecules in order to synthesize ATP. Aerobic respiration, which involves the participation of oxygen, consists of four distinct but linked stages, glycolysis, the link reaction, Krebs cycle and the electron transport chain.

Glycolysis

This takes place in the cytoplasm and oxygen is not required.

→ The glucose molecule is first phosphorylated to make it more reactive by the addition of two molecules of ATP to form hexose phosphate.

→ The 6-carbon hexose phosphate is split into two molecules of triose phosphate (3-carbon sugars).

→ Each 3-carbon sugar is converted to pyruvic acid.

→ Two of the steps transfer sufficient energy for the synthesis of ATP yielding a total of four ATP molecules, and since these are formed directly from phosphate compounds in the process it is known as substrate-level phosphorylation.

→ Each 3-carbon sugar initially donates two hydrogen atoms to reduce the carrier molecule NAD. If oxygen is available, NADH/H$^+$ enters the mitochondrion and brings about the synthesis of 6 ATP molecules.

→ Thus, the breakdown of one molecule of glucose produces two molecules of ATP, two molecules of reduced NAD and two molecules of pyruvate.

Link reaction

→ Pyruvic acid diffuses from the cytoplasm to the mitochondrial matrix.

→ The two molecules of pyruvate (3-carbon) from glycolysis are converted to two molecules of 2-carbon acetate with the formation of two molecules of reduced NAD and the loss of two molecules of carbon dioxide.

→ The acetate then combines with coenzyme A to form **acetyl coenzyme A.**

Krebs cycle

→ The acetyl CoA then enters the Krebs cycle by combining with a 4-carbon acid to form a 6-carbon compound, citric acid; the CoA is regenerated.

→ The 6-carbon compound undergoes reactions during which carbon dioxide and hydrogen atoms are removed. After the acetate fragment is broken down, the remaining 4-carbon residue undergoes conversion to regenerate the 4-carbon compound which combines with more acetyl CoA.

→ Two of the steps involve **decarboxylation**, and four of the steps involve **dehydrogenation**.

→ The hydrogen atoms produced are collected by two different carriers with the formation of three molecules of NADH/H$^+$ (reduced NAD) and one molecule of FADH$_2$ (reduced FAD).

→ Thus, for each turn of the cycle the overall production is one ATP, three reduced NAD and one reduced FAD. Since these are produced from the breakdown of one molecule of pyruvate, the two molecules produced in glycolysis will give double this yield.

Checkpoint 1

How much energy is released from one molecule of glucose?

Examiner's secrets

You are not required to know the names of all the enzymes involved at each stage of glycolysis. However, you should know that dehydrogenase enzymes catalyze the removal of hydrogen atoms and decarboxylase catalyzes the removal of carbon dioxide.

Checkpoint 2

What is the net gain of ATP molecules in glycolysis?

Watch out!

For each glucose molecule broken down, the cycle turns twice. So far we have six NADH/H$^+$, two FADH$_2$ and two ATPs produced by substrate-level phosphorylation.

Checkpoint 3

What is the name of the enzyme which catalyzes the removal of carbon dioxide?

Electron transport system

NAD and FAD are both **coenzymes** and are the hydrogen acceptor molecules. They pass on the hydrogen atoms through a series of carriers known as the electron transport chain and generate ATP by the proton pump mechanism.

For the breakdown of one molecule of glucose:

Glycolysis	Link reaction	Krebs Cycle
2 NADH + 2 ATP directly (net)	2 NADH	6 NADH + 2 ATP directly 2 $FADH_2$

Ten molecules of NADH are made and each molecule produces three ATP molecules giving 30 ATP.

Two molecules of $FADH_2$ are made and each molecule produces two ATP molecules giving 4 ATP.

Four ATP are made directly (substrate-level phosphorylation).

This gives a total of 38 molecules of ATP produced from each molecule of glucose.

Links

See also pages 86 and 87 on ATP synthesis.

Checkpoint 4

Explain the importance of the oxidoreductase enzymes, dehydrogenase and oxidase.

Exam practice
answers: page 98

1 The diagram below represents an outline of stages involved in respiration in a plant cell.

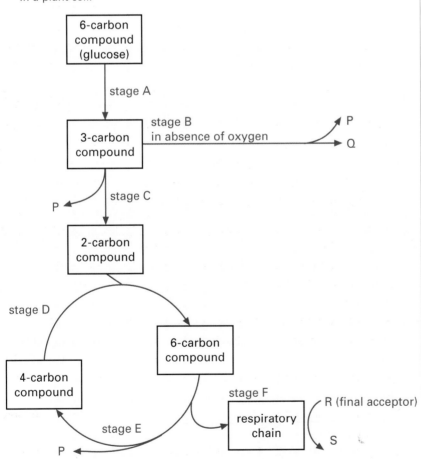

Action point

Draw a diagram that summarizes aerobic respiration and insert at the appropriate points where, and how many, ATP molecules are formed.

(a) Name the compounds represented by P, Q, R and S.

(b) (i) Which letter represents the stage during which most ATP is produced?

  (ii) Which two letters represent separate stages where ATP is not produced?

(c) Indicate the nature of the reactions in stage F.s

Anaerobic respiration and energy budgets

Anaerobic respiration occurs in some microorganisms and in some tissues, e.g. muscle cells, which may be temporarily deprived of oxygen. However, without oxygen, oxidative phosphorylation will not take place and so the yield of ATP molecules will be greatly reduced. You need to be able to compare and explain the differences in yields of ATP from aerobic and anaerobic respiration.

Anaerobic respiration

Without oxygen the carriers reduced NAD and reduced FAD cannot be reoxidized and therefore made available to pick up more hydrogen, and so the link reaction and the Krebs cycle cannot take place. Only glycolysis can occur in the absence of oxygen.

→ In vertebrate muscle cells **lactic acid** is produced. The pyruvate itself becomes the hydrogen acceptor.

→ In higher plants and yeast, **alcohol** is the product. This is also known as alcoholic fermentation. Ethanal (acetaldehyde), formed by the decarboxylation of pyruvate, is the hydrogen acceptor.

A considerable amount of energy remains in the ethanol and lactic acid.

Energy budget

Aerobic respiration
Glycolysis

2 ATP used (in the phosphorylation of glucose)
4 ATP produced directly (by substrate-level phosphorylation)
6 produced (from two pairs of hydrogen atoms picked up by NAD which may enter the mitochondrion and pass along the electron transport system in aerobic conditions)
Net gain from glycolysis = **8 ATP** (if aerobic), **2 ATP** (if anaerobic)

Link reaction

3 ATP (2 ×) = **6 ATP**

Krebs cycle

2 × 1 ATP produced directly (by substrate-level phosphorylation)
2 × 11 ATP from electron transport chain
(three pairs of hydrogen atoms via NAD yields 9 ATP; one pair via FAD yields 2 ATP)
Net gain from Krebs cycle = **24 ATP**
Total net gain = **38 ATP**

The efficiency of aerobic respiration

Each ATP molecule will yield 30.7 kJ of energy
Therefore total energy available is $38 \times 30.7 = 1\,166.6$ kJ
The complete oxidation of glucose yields 2 880 kJ

Therefore efficiency $= \dfrac{38 \times 30.7 \times 100}{2\,880} = 40.5\%$

(not all the energy of the glucose molecule is captured in ATP as there is a loss of energy as heat)

Checkpoint 1

Under what temporary conditions do plants respire anaerobically?

Checkpoint 2

Why is yeast unable to respire anaerobically for an indefinite period?

Checkpoint 3

How can the energy locked up in lactic acid be released by muscle cells?

Watch out!

A total of 38 ATPs is the value most commonly quoted as being produced during aerobic respiration. In fact, 2 ATP are (usually) used up in transporting some of the metabolites across the inner mitochondrial membrane, so the value of 36 is given in some texts. As long as you know why there are these two values, either is acceptable.

Checkpoint 4

What is the efficiency of a car engine?

Anaerobic respiration

Energy yield = 2 ATP (glycolysis only)

Therefore total energy available = 61.4 kJ

$$\text{Efficiency} = \frac{61.4 \times 100}{2\,880} = 2.13\%$$

Alternative respiratory substrates

In addition to glucose both fats and proteins may, in certain circumstances, be used as respiratory substrates.

Fat respiration

Fat is a good energy store and is used as a respiratory substrate when carbohydrate is in short supply. It has to be split into its constituent molecules of **glycerol and fatty acids**, first by hydrolysis and then the glycerol is converted into a 3-carbon sugar which enters the Krebs cycle via triose phosphate, yielding ATP molecules. The long fatty acid molecules are split into 2-carbon fragments which enter the pathway as acetyl CoA. Very large numbers of ATP molecules are built up during the process, the precise number depending on the length of the hydrocarbon chain of the fatty acid, e.g. the complete oxidation of stearic acid yields nearly four times as many ATP molecules as given by the oxidation of a single glucose molecule.

Protein

Protein is very rarely used as a respiratory substrate, usually only when all reserves of carbohydrate and fat have been used up. The protein is hydrolysed into its constituent **amino acids** and then is deaminated in the liver. The amino group is converted into urea and excreted and the residue is converted to either acetyl CoA, pyruvic acid or some other Krebs cycle intermediate and oxidized.

Respiratory quotient (RQ)

It is possible to work out the respiratory quotient for a particular organism if the volume of carbon dioxide produced and the volume of oxygen taken up in a given period of time is known.

$$RQ = \frac{\text{volume of carbon dioxide produced}}{\text{volume of oxygen used}}$$

This information can give an indication of the type of food being used as the respiratory substrate by working out the theoretical RQ values for carbohydrates, fat and protein, and then comparing them with actual values obtained in experiments using respirometers. Carbohydrates give an RQ of 1.0, fats about 0.7 and proteins 0.9. The human RQ is usually about 0.85 indicating that a mixture of carbohydrates and fat is used for respiration.

> **Action point**
>
> Construct a simplified diagram of the respiratory pathway showing where fat and protein (as glycerol and fatty acid, and amino acids) enter the pathway.

> **Grade booster**
>
> Research the effect of exercise on the rate of metabolism.

> **Checkpoint 5**
>
> Calculate the RQ if 26 molecules of oxygen are required to completely oxidize stearic acid (a fat) and 18 molecules of carbon dioxide are evolved.

> **Checkpoint 6**
>
> Would animals living in conditions of low oxygen have a high or low RQ value?

> **Action point**
>
> Draw a diagram of a simple respirometer to show the rate of oxygen consumption in a small organism such as a locust.

Exam practice   answers: page 98

1 From your knowledge of the electron transport chain explain fully why glucose is not broken down completely during anaerobic respiration.

Photosynthesis

Photosynthesis takes place in the chloroplasts, which are found in the mesophyll cells and guard cells of green leaves. In the chloroplasts the energy of sunlight is trapped by the pigment chlorophyll. Since photosynthesis is a biochemical process involving many enzymes, a suitable temperature is necessary, in addition to light, carbon dioxide and water.

Grade booster

Research how commercial growers apply a knowledge of limiting factors in enhancing conditions in glass houses.

Factors affecting photosynthesis

The rate at which oxygen is given off by green plants can be used to measure the rate of photosynthesis. Blackman, in his experiments in the 1930s, showed that bubbles of oxygen were given off from the cut end of a stem of an aquatic plant such as *Elodea* when it was immersed in illuminated pond water. He showed that photosynthesis is controlled by a combination of factors and the rate is limited by whatever factor is nearest its minimum value. He called this the law of limiting factors.

Carbon dioxide

The concentration of carbon dioxide in the atmosphere is about 0.03% and on a warm, sunny day when the light intensity and temperature are both high it can be the factor that limits the rate of photosynthesis.

Light intensity

Increasing the light intensity causes the rate of photosynthesis to increase, up to a critical point, at which some other factor becomes limiting. Above this point, any increase in the light intensity will have no further effect on the rate.

Checkpoint 1

Suggest how plants living in low light intensity (shade plants) benefit from having shorter compensation periods than sun plants.

Compensation point

For a plant to grow it has to synthesize organic materials more rapidly than it oxidizes them in respiration. When photosynthesis and respiration occur at rates such that there is no gain or loss of organic matter, a plant is at its compensation point. At this point there is no net exchange of gases. The time taken for a plant that has been in darkness to reach the compensation point is called the compensation period.

Action point

Draw a graph showing the effect of light intensity on the rate of photosynthesis as measured by the amount of carbon dioxide exchanged over a 24-hour period.

Temperature

Since photosynthesis is a biochemical process the enzyme-catalyzed reactions are affected by temperature.

Action point

Draw graphs showing the effects of light intensity, carbon dioxide concentration and temperature on the rate of photosynthesis.

Water

Water is necessary for a great many metabolic reactions taking place in the plant. As a substrate in the process of photosynthesis its deficiency will reduce the rate of photosynthesis.

Photosynthetic pigments

The role of the photosynthetic pigments is to absorb light energy and to convert it to chemical energy. There are two types of pigments in flowering plants.

→ The **chlorophylls** absorb mainly in the red and blue-violet regions of the spectrum. There are a number of different chlorophylls, with chlorophyll *a* and *b* being the most common.
→ The **carotenoids** absorb blue-violet light. They are thought to act as accessory pigments. There are two main types of carotenoids, the carotenes and the xanthophylls.

Absorption and action spectra

Chlorophyll absorbs only certain wavelengths, reflecting others. An absorption spectrum indicates that chlorophyll absorbs wavelengths in the blue and red parts of the spectrum. However, this does not indicate whether the wavelengths are actually used. This can be shown by plotting an action spectrum of the amount of carbohydrate synthesized by plants exposed to different wavelengths of light. Since the absorption spectrum and the action spectrum for chlorophyll show a close correlation it suggests that this pigment is responsible for absorbing the light used in photosynthesis.

Leaf structure and function

Before considering the biochemistry of photosynthesis it is necessary to consider how well the structure of the leaves of flowering plants is related to their function:

→ large surface area to capture as much sunlight as possible
→ thin, so that light can penetrate to the lower layers and there is a short diffusion path for carbon dioxide
→ the cuticle and epidermis are transparent, to allow light through
→ the palisade mesophyll cells are packed with chloroplasts and arranged with their long axes perpendicular to the surface
→ chloroplasts can move intracellularly, a process called cyclosis, allowing them to arrange themselves into the best position within the cell for the efficient absorption of light
→ chloroplasts hold the chlorophyll in an ordered arrangement
→ stomata allow carbon dioxide entry
→ intracellular spaces

Checkpoint 2

Why are leaves green?

Checkpoint 3

What technique is used to demonstrate the presence of the different pigments in green leaves?

Checkpoint 4

Why is it an advantage for a leaf to contain more than one pigment?

Checkpoint 5

Why do leaves appear red, orange and yellow in the autumn?

Watch out!

Don't confuse the absorption spectrum and action spectrum. The action spectrum shows the actual wavelengths that are used in photosynthesis.

Action point

Draw the action spectrum for photosynthesis and the absorption spectrum for chlorophyll on the same graph.

Checkpoint 6

What is the function of the leaf vein?

Exam practice   answers: pages 98, 99

1   Describe how the structure and distribution of chloroplasts ensure the efficient trapping of light by leaves.

2   Give an illustrated account of the anatomy of a typical leaf. State the functions of the structures named.

Action point

Describe in your own words how the leaf is adapted for photosynthesis by rephrasing the bullet lists as an essay-style paragraph.

Biochemistry of the light-dependent stage

Photosynthesis is a three-stage process. You will first need to consider how light energy is captured by the plant. The light-dependent reactions result in the formation of ATP and reduced NADP, which are then used in the light-independent stage in the reduction of carbon dioxide and the synthesis of organic compounds.

Watch out!

Textbooks differ in their use of the term 'reduced NADP'. Some use $NADPH_2$, others $NADPH^+H^+$ (the correct terminology used by the Biochemical Society in their booklet 'Biochemistry across the school curriculum'). Throughout this section the term 'reduced NADP' will be used.

Light harvesting

Light energy is captured by the plant using a mixture of pigments including chlorophyll. Within the thylakoid membranes of the chloroplast, chlorophyll molecules are arranged along with their accessory pigments into groups of several hundred molecules. Each group is called an antenna complex. Special proteins associated with these pigments help to funnel photons of light energy entering the chloroplast. Chlorophyll *b* is an accessory pigment molecule that passes on its energy to chlorophyll *a*, which is a primary pigment molecule and is known as the reaction centre.

There are two types of photosynthetic units, with *Photosystem I (PSI)* at a higher energy level than *Photosystem II (PSII)*.

The jargon

Radiant energy comes in discrete packets called *quanta*. A single quantum of light is called a *photon*.

The jargon

Photosystems are groups of chemicals that harness the light and pass on energy.

Light-dependent stage

During the light-dependent stage electron transport occurs as follows. The absorption of light energy by the photosynthetic pigments causes the displacement of 'excited' (high-energy) electrons which are accepted by other molecules called electron acceptors. This is an oxidation–reduction process in which the pigment molecule, the electron donor, is oxidized and the acceptor molecule is reduced. As electron transport occurs hydrogen ions are pumped from the stroma into the thylakoid discs which make up each granum. These H^+ flow out and down a gradient of both concentration and electrical potential difference with the production of ATP (see diagram below).

Checkpoint 1

What is the source of electrons for Photosystem II?

Checkpoint 2

Which enzyme is involved in ATP synthesis?

Grade booster

Compare ATP production in mitochondrion and chloroplast.

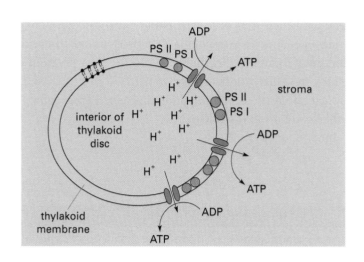

The light-dependent stage occurs in the thylakoids of the chloroplasts and involves the splitting of water by light (photolysis). In the process ADP is converted to ATP. This is known as **photophosphorylation** (see diagram below).

Electrons from chlorophyll are passed to the light-independent reaction via reduced NADP. These electrons are replaced by others from the water molecule. Since these electrons are *not* recycled back into the chlorophyll this method of ATP production is called **non-cyclic** photophosphorylation. The process occurs as follows.

→ Light energy absorbed by PSII causes the displacement of an excited electron to a higher energy level. The electrons are received by an electron acceptor and passed via a carrier through a proton pump, from which the electrons are carried to PSI. The energy lost by the electrons is captured by converting ADP to ATP.

→ Light energy absorbed by PSI boosts the electrons to an even higher energy level and are picked up by another electron acceptor. The electrons which have been removed from the chlorophyll are replaced by the electrons from a water molecule that dissociates into protons and oxygen. The second electron acceptor, which receives electrons from PSI, passes some of these electrons to hydrogen ions on the outside of the thylakoid membrane, where they reduce NADP.

ATP can also be generated by **cyclic** photophosphorylation. Some electrons from PSI return to the chlorophyll directly via the electron carrier system, forming ATP in the process. These electrons are recycled, harnessing energy from light and generating ATP. No reduced NADP is produced.

Checkpoint 3

With reference to the build-up of protons in the mitochondrion and chloroplast, explain how oxidative phosphorylation differs from photophosphorylation.

Action point

Restructure the description of non-cyclic photophosphorylation as a list of bullet points.

Checkpoint 4

Make a table of differences between cyclic and non-cyclic photophosphorylation.

Checkpoint 5

What are the three products of the light-dependent stage?

Examiner's secrets

In examinations this diagram may be reproduced whole or in parts. Often you are required to fill in empty boxes where key substances have been omitted.

Exam practice   answers: page 99

1 The synthesis of ATP is driven by a flow of protons. Describe where and how the required proton gradient is produced in the chloroplast.

Watch out!

Electrical neutrality of PSII is achieved using electrons donated from the splitting of water.

Biochemistry of the light-independent stage

The light-independent stage occurs in the stroma of the chloroplast and involves many reactions each catalyzed by a different enzyme. The reactions use the products of the light-dependent stage, ATP as a source of energy, and reduced NADP as the source of the reducing power to reduce carbon dioxide and synthesize hexose sugar.

The Calvin cycle

The jargon

Rubisco is an abbreviation for the enzyme ribulose bisphosphate carboxylase oxygenase!

The sequence of events in this stage of photosynthesis was worked out by Calvin and his associates using ^{14}C, a radioisotope of carbon, and the unicellular algae *Chlorella*.

→ A 5-carbon acceptor molecule, ribulose bisphosphate (RuBP), combines with carbon dioxide (catalyzed by Rubisco) forming an unstable 6-carbon compound.

→ The 6-carbon compound immediately splits into two molecules of a 3-carbon compound called glycerate-3-phosphate (GP).

Watch out!

GP is also known as phosphoglyceric acid (PGA).

→ GP is phosphorylated by ATP and then reduced by reduced NADP (from the light-dependent stage) to glyceraldehyde-3-phosphate (GALP).

→ Some of this 3-carbon sugar can be built up into glucose phosphate and then into starch by condensation.

Action point

Describe the Calvin cycle by rephrasing the bullet lists.

→ In order that the cycle continues, the majority of the GALP formed enters a series of reactions driven by ATP which results in the regeneration of RuBP (see diagram below).

Checkpoint 1

Why is it important that RuBP is regenerated?

Checkpoint 2

How many molecules of the 3-carbon triose phosphate are needed to regenerate three molecules of RuBP?

Watch out!

Once supplies of ATP and reduced NADP are used up then photosynthesis ends. The light-independent reaction can continue for a short time until stores of ATP, reduced NADP and GP are exhausted.

Products of photosynthesis

The plant must synthesize all organic materials from the intermediates and products of photosynthesis. The diagram below shows an outline of the pathways involved.

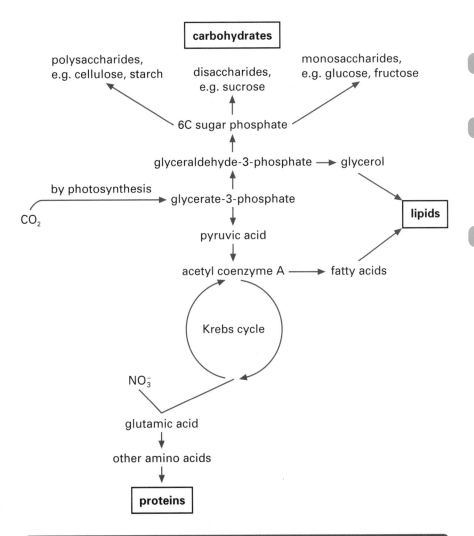

Checkpoint 3

Name a polymer of glucose.

Checkpoint 4

What must glycerate-3-phosphate be converted to before it can enter the Krebs cycle and become part of the process of amino acid synthesis?

Checkpoint 5

Which one intermediate of photosynthesis and one intermediate of respiration is needed to produce lipids?

Mineral nutrition

Mineral ions are required for the synthesis of compounds needed for the growth of the plant. Macronutrients, e.g. potassium, sodium, magnesium, calcium, nitrogen and phosphorus, are required in substantial quantities but the micronutrients, e.g. manganese and copper, are needed in much smaller amounts.

→ Nitrogen is transported as nitrates in the xylem and as amino acids in the phloem. It is needed for amino acid synthesis for proteins and nucleic acids.
→ Phosphorus is transported as inorganic phosphate ions in the xylem and as sugar phosphate in the phloem for ATP, nucleic acid and phospholipid production.
→ Magnesium ions are transported in the xylem and are used to make chlorophyll.

Checkpoint 6

Why do plants deficient in magnesium tend to have very yellow leaves?

Links

Recall the nitrogen cycle on page 74 and the uptake of minerals by roots on page 54.

Examiner's secrets

You only need to learn about the three minerals described in the text.

Exam practice   answers: page 99

1 Describe the sequence of reactions by which carbohydrate is produced in the light-independent stage of photosynthesis, indicating where in the cell these reactions take place.

2 Show how the plant utilizes carbohydrates from the dark stage to make other products. Indicate the function of these products.

Answers
Metabolism, respiration and photosynthesis

The role of ATP in metabolism

Checkpoints

1 ATP => ADP + Pi + energy.
2 Protein synthesis, DNA replication, polysaccharide from monosaccharide, etc.
3 Increased surface area for enzyme reactions.
4 No oxygen to accept hydrogen atoms so they build up.
5 Oxidative phosphorylation.

Exam questions

1 Your answer should contain the following points in a logical sequence:
   - The energy of electrons is used to drive proton pumps.
   - The electrons are obtained from the oxidation of reduced NAD, which drives three pumps, or from the oxidation of reduced FAD, which drives two pumps.
   - Electrons are conveyed from pump to pump by carriers.
   - In order to function, pumps and carriers must be assembled in proximity to one another.
   - The membrane forming the cristae performs this function.
   - The enzymes of the Krebs cycle are situated in the matrix.
   - The enzymes provide most of the reduced NAD (and FAD).
   - Oxygen is the acceptor of the final low-energy electrons, with the formation of water.
   - The pumps create a proton gradient across the membrane of the cristae.
   - There is a high concentration of protons in the intermembranal space and a low concentration in the matrix.
   - ATP is formed by the enzyme ATPase.
   - ATPase is a transmembrane protein in the cristae.
   - The flow of protons down the gradient through the enzyme causes the synthesis of ATP and its release into the matrix.

Aerobic respiration

Checkpoints

1 2800 kJ.
2 2 (input of 2, 4 ATP produced).
3 Decarboxylase.
4 The redox reactions: when a substance is oxidized it loses electrons and when it is reduced it gains electrons; in this way electrons pass along from carrier to carrier.

Dehydrogenase catalyzes the removal of H+ from a substrate and makes it available to be taken up by the electron transport system. Oxidase removes H+ at the end of the electron transport chain.

Exam practice

1 (a) P, carbon dioxide; Q, alcohol (ethanol); R, oxygen; S, water.
   (b) (i) F. (ii) A and D.
   (c) Oxidation–reduction reactions.

Anaerobic respiration and energy budgets

Checkpoints

1 Roots in waterlogged soil.
2 Its product, ethanol, reaches a toxic level.
3 Lactic acid is transported to the liver and converted back to pyruvate; some is oxidized to carbon dioxide and water and the rest is converted to glucose and stored as glycogen.
4 25%.
5 RQ = 0.69.
6 High RQ.

Exam practice

1 There is no oxygen available to remove H+ ions/there is no oxygen to remove electrons. When the electrons come off the carrier system there is no final electron acceptor and so water cannot be formed.

Photosynthesis

Checkpoints

1 So that shade plants can make more efficient use of light at low intensities.
2 The light in the green part of the spectrum is not absorbed but is reflected.
3 Chromatography.
4 To increase the range of wavelengths from which energy may be obtained.
5 Chlorophyll is the pigment that is broken down prior to leaf fall exposing the colours of the carotenoids.
6 The xylem in the vein transports water and the phloem transports sucrose.

Exam practice

1 Your answer should include the following points:
   - The chloroplast is surrounded by a double membrane, enclosing the stroma.

- Within the stroma are the thylakoid membranes, which enclose the thylakoid space.
- The photosystems are held on the thylakoid membranes.
- Photons of light energy produce high energy electrons in the photosystems.
- The efficient trapping of light requires large numbers of these reaction centres.
- Thus the area of the thylakoid membranes must be large.
- Its area is increased by being stacked into grana.
- Chloroplasts occur in all the mesophyll cells of the leaf but are more abundant in the palisade cells than in the spongy mesophyll cells.
- Palisade cells are uppermost in the leaf, closest to the light.
- They are densely packed cylinders with a long axis aligned to the light direction.
- Chloroplasts can migrate up and down the cylinder to obtain optimum exposure to light.

2 Your diagram should be well annotated (see diagram below).

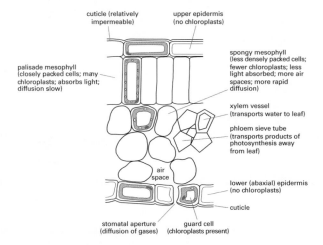

Biochemistry of the light-dependent stage

Checkpoints

1 OH- ions from the splitting of water.

2 ATPase.

3 In the mitochondrion the build-up of protons is outside the mitochondrial membrane whereas in the chloroplast the build-up of protons is inside the thylakoid disc.

4 Cyclic: ATP produced, PSI involved only.
Non-cyclic: ATP produced, PSI and PSII involved, reduced NADP produced.

5 Reduced NADP, ATP and oxygen.

Exam practice

1 Your answer should include 10 of the following points:
- The proton gradient is across the thylakoid membrane.
- There is a high concentration in the thylakoid space and low in the stroma.
- Protons are pumped from the stroma to the thylakoid space.
- The energy for the pumps comes from the excited electrons released from the photosystems by photons.
- The electrons are passed to and from the pump by a chain of carriers.
- Some of the electrons from PSI return as low-energy electrons – the cyclic system.
- Some PSI electrons combine with protons to reduce NADP to $NADPH_2$.
- This takes place on the stroma side of the membrane and the removal of stroma protons increases the gradient.
- Lower-energy electrons from PSII pass along the electron transport chain to PSI.
- This pumps more protons in the process.
- The electrons lost from PSII are replaced from the splitting of water.
- This takes place on the thylakoid space side of the membrane.
- This adds protons to the gradient.

Biochemistry of the light-independent stage

Checkpoints

1 So that carbon dioxide can be fixed and the cycle can continue.

2 Five.

3 Starch, cellulose.

4 Acetyl CoA.

5 Glycerate-3-phosphate and acetyl CoA.

6 Magnesium is needed to make the green pigment, chlorophyll.

Exam practice

1 Your answer should include 10 of the following points:
- carbon dioxide is combined with a 5-carbon molecule called ribulose bisphosphate
- this forms an unstable 6-carbon compound which immediately breaks down to two 3-carbon molecules called glycerate-3-phosphate (or GP/PGA)
- this is phosphorylated (with ATP)
- it then reacts with reduced NADP
- this is a reduction reaction
- glyceraldehyde-3-phosphate (GALP) is formed
- GALP reacts with ATP
- to re-form ribulose bisphosphate
- or it can be converted into hexose sugar
- the reactions take place in the stroma of the chloroplast
- the stroma is the medium filling the space between the thylakoids and the outer membranes of the organelle

2 Your answer should include the following points:
- glucose is used as an energy source in cellular respiration
- it can be converted to other monosaccharides
- and combined to give disaccharides, particularly sucrose
- sucrose is the main transport molecule in phloem
- it can be synthesized into:
  → polysaccharides
  → starch for energy store
  → cellulose for cell wall formation
- sugars can be converted to amino acids
- and built into plant protein
- carbohydrates can be metabolized to lipids
- use of lipids – oil storage in seeds, phospholipids in membranes, etc.

Revision checklist
Metabolism, respiration and photosynthesis

By the end of this chapter you should be able to:

1	Outline glycolysis, the link reaction and the Krebs cycle, indicating the points where hydrogen atoms and carbon dioxide are evolved and where ATP is produced.	Confident	Not confident **Revise** page 88
2	Explain the process of chemiosmosis with reference to the electron transport chain, protein gradients and ATP synthetase.	Confident	Not confident **Revise** page 86, 87
3	Define the term 'limiting factor' and explain how carbon dioxide concentration, light intensity and temperature affect the rate of photosynthesis.	Confident	Not confident **Revise** page 92
4	Explain the importance of photosynthetic pigments in photosynthesis.	Confident	Not confident **Revise** page 93
5	Outline how light energy is converted to chemical energy in the light-dependent stage of photosynthesis.	Confident	Not confident **Revise** pages 94, 95
6	Explain how the products of the light-dependent stage (ATP and reduced NADP) are used in the Calvin cycle.	Confident	Not confident **Revise** page 96

Regulation and control

To survive, plants and animals must react to changes in their external environment. They must have mechanisms for detecting such changes and bringing about appropriate responses. Since the structures that detect changes may be some distance from those that respond, a means of communication within the body is needed. Animals have nervous systems and hormonal mechanisms, whereas plants have only the latter. Environmental stimuli must be monitored and the information fed into the appropriate system. Both plants and animals respond to the information transmitted.

Exam themes

- The principles of homeostasis in terms of receptors, effectors and negative feedback
- How blood glucose concentration is regulated by negative feedback control mechanisms
- Thermoregulation in mammals
- The nature of communication systems in flowering plants
- The commercial application of plant growth regulators
- The transmission of an axon potential in a myelinated nerve
- The organization of the nervous system with reference to the central and peripheral systems
- The structure and functions of the mammalian brain
- The structure and function of a cholinergic synapse
- The role of sensory receptors in mammals in converting different forms of energy into nerve impulses
- The structure of the eye and the function of its parts
- The retina of the eye with reference to visual acuity, colour vision and sensitivity to different light intensities
- The histology and ultrastructure of striated muscle
- The sliding filament theory of muscle contraction

Topic checklist

	Edexcel		AQA		OCR		WJEC	
	AS	A2	AS	A2	AS	A2	AS	A2
Homeostasis		●		●		●		●
The kidney		●		●		●		●
Functioning of the kidney		●		●		●		●
Plant growth substances		●		●		●		
Effect of light on plant growth		●		●		●		●
Nerves		●		●		●		●
Synapses		●		●		●		●
The central nervous system		●		●		●		●
Sensory receptors		●		●		●		●
Muscles		●		●		●		

Homeostasis

The term 'homeostasis' is used to describe the mechanisms by which a constant internal environment is achieved. An example of the advantage of such constancy is the greater environmental freedom achieved by homoiothermic animals. The control of blood sugar is another example of homeostasis where the supply of glucose to the metabolizing cells must be kept constant even though feeding is intermittent.

"The constancy of the internal environment is the condition of the free life."

Claude Bernard, 19th century French biologist

The homeostatic process

All homeostatic processes involve a **detector** (or sensor) that monitors the factor being controlled. When the detector senses a change from the normal, it informs a **controller**, a coordinating system that decides upon an appropriate method of correcting the deviation. The controller communicates with one or more **effectors**, which carry out the corrective procedures. Once the correction is made and the factor returned to normal, information is fed back to the detector which then 'switches off' (see diagram below). This is what happens in most biological control systems, i.e. the controller is no longer alerted to the deviation from the normal. This is called negative feedback.

The jargon

An *endocrine* or *ductless* gland has no duct to deliver its secretions so the secretions are released into the blood as it flows through the gland.

Hormones

The endocrine system consists of a number of chemicals called hormones that regulate the body's activities.

Hormones are:

→ produced in the tissues of endocrine glands and released into the bloodstream
→ carried in the blood plasma to a target organ where they trigger a response. Most do this by activating enzymes

Checkpoint 1

How would abnormally high concentrations of glucose in the blood affect the water content of the body?

Regulation of blood sugar

Cells are sensitive to changes in sugar level and it is therefore important to maintain 90 mg of glucose in each 100 cm^3 of blood. The supply of carbohydrates in mammals fluctuates because they do not eat continuously throughout the day and the quantity of carbohydrate varies from meal to meal. As cells metabolize continuously they need to be provided with a constant supply of glucose.

The liver plays a key role in glucose homeostasis. If the glucose concentration in the blood is higher than normal, the excess is stored as glycogen in the liver. If the level is less than normal, glycogen reserves are converted to glucose. Patches of pancreatic cells, called the islets of Langerhans, control this mechanism. There are two types of cells in the islets:

Watch out!

It is not the hormones which directly interconvert glucose and glycogen. Hormones control enzymes which catalyze the conversions.

Links

The potential use of stem cells to treat diabetes mellitus is considered on page 180.

→ If applied to dwarf varieties of plants, they stimulate the rapid growth of the stem and the plant develops to normal size by increasing internode growth.

→ They are involved in the breaking of dormancy in buds and seeds. Studies with cereal grains have shown that the release of enzymes takes place from the aleurone layer surrounding the endosperm. After the absorption of water by the grain there is a marked rise in embryonic production of gibbcrelins, which diffuse into the endosperm, where food reserves are hydrolyzed, so providing the embryo with materials for growth. Gibberelins and auxins supplement each other in stem elongation.

Gibberelins are used commercially to promote the setting of fruit and in parthenocarpy.

Cytokinins

The cytokinins form a group of plant growth substances that are involved in cell division. Together with auxin, they promote cell division in the apical and lateral meristems. They have also been shown to cause delay in the ageing of leaves, and are involved in the breaking of dormancy in buds and leaves.

Abscisic acid

Abscisic acid is a growth inhibitor and appears to be antagonistic to the action of the other growth substances. Its main effect is on abscission (leaf and fruit fall). The process results from a balance between the production of auxin and abscisic acid. As the fruit ripens the level of auxin (which inhibits abscission) falls, while that of abscisic acid (which promotes abscission) increases. This leads to the formation of an abscission layer, which causes the fruit to fall.

Ethene

Ethene is a gas that is produced by most plant organs. It appears to stimulate the rapid increase in respiration rate that occurs before the ripening of fruit. Knowledge of its effects has been used in the commercial control of ripening. Fruit is stored in oxygen-free conditions to prevent ripening, which can be induced by applying oxygen and ethene when required.

The jargon

An *internode* is the part of the shoot between one bud and the next.

Links

See germination on pages 154, 155.

Checkpoint 4

Why are some synthetic growth regulators more useful to us than naturally occurring substances?

Checkpoint 5

If you removed the apical bud from a shoot and applied cytokinin to the cut surface what would you expect to happen a few days later?

Watch out!

Be aware that growth substances interact with one another either by supplementing each other (synergism) or by having an antagonistic effect.

Checkpoint 6

Why does a ripe banana, placed near unripe fruit but not in contact with them, speed up their ripening?

Grade booster

Research the commercial applications of plant hormones.

Exam practice answers: page 123

1 Explain why auxins, cytokinins and gibberelins can be important to gardeners and farmers.

Effect of light on plant growth

Light affects many aspects of plant growth, from the synthesis of chlorophyll to photosynthesis and phototropic movements. A number of plant growth responses are influenced differently by light of different wavelengths. For light to have an effect it must be absorbed by a photoreceptor substance, phytochrome. Flowering is regulated by day-length.

Checkpoint 1

In experiments on phototropism the coleoptiles are grown in the dark except for brief periods of exposure to unilateral light. Why is it necessary to keep coleoptiles in the dark?

Checkpoint 2

Suggest how the response of a shoot to light is important for successful growth of a seedling.

Checkpoint 3

If a shoot tip is covered with tinfoil and light is shone unilaterally what will be the response?

Checkpoint 4

Coleoptiles subjected to unilateral light have their tips removed and placed on agar blocks for several hours. When the blocks are placed on the tops of decapitated coleoptiles but on the right side only, in which direction will the coleoptiles bend?

Phototropism

A **tropism** is a growth movement of part of a plant in response to a directional stimulus. The response to light is termed phototropism. The direction of the response is described as positive if movement is towards the stimulus, or negative if it is away from it.

Early experiments on auxins used oat coleoptiles.

→ If oat coleoptiles were subjected to unilateral illumination, they showed a positive phototropic response and grew towards the direction of the light, the bending occurring just behind the tip.

→ If the tip of the coleoptile was removed, then no response was shown to the direction of the light, and similarly no response was shown if the tips were covered.

→ If the coleoptile was covered except for the tip, then a positive phototropic response was again shown.

As a result of these experiments, it could be concluded that the tip detected the stimulus of light and that the response was made in the region just behind the tip (see diagram below).

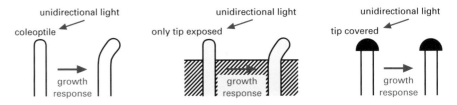

Phytochrome

The photoreceptor responsible for absorbing light has been identified as **phytochrome**, a blue-green pigment found in very minute quantities in plants. Phytochrome exists in two forms that are interconvertible:

→ one form absorbs red light, with an absorption peak at 660 nm, and is referred to as Pr

→ the second form absorbs far-red light, with an absorption peak at 730 nm, and is referred to as Pfr

When Pr absorbs red light it is converted to Pfr. There is more red light than far-red light in natural sunlight, so Pfr tends to accumulate during the day and become slowly converted back to Pr at night. It has been shown that the phytochrome system is important in many aspects of plant growth, such as germination in some seeds, stem elongation, leaf expansion, growth of lateral roots and flowering.

Checkpoint 5

Why does Pfr tend to accumulate during the day?

Photoperiodism

Photoperiodism is the term used to describe the influence of the relative lengths of periods of light and darkness on the activities of plants, and day-length has been shown to have an important effect on flowering. Flowering plants can be divided into three groups according to their photoperiodic requirements prior to the production of flowers.

→ *Day-neutral plants*: flowering does not seem to be affected by day-length, e.g. tomato, cucumber, antirrhinum.
→ *Long-day plants*: flowering is induced by exposure to dark periods shorter than a critical length, e.g. petunia, spinach.
→ *Short-day plants*: flowering is induced by exposure to dark periods longer than a critical length, e.g. chrysanthemum, tobacco, poinsettia.

Flowering in short-day plants is inhibited by exposure to red light, while exposure to far-red light will bring about flowering. It seems that these plants will flower only if the level of Pfr is low enough. The situation in long-day plants is reversed and flowering is triggered by high levels of Pfr. The length of the photoperiod is less critical than the length of the dark period, and if the photoperiod is interrupted with a short period of darkness flowering still follows (see diagram below). If the dark period is interrupted by as little as 1-min exposure to light, flowering is prevented. Red light is most effective in this respect yet the effect of red light treatment can be overcome if the plant is immediately exposed to infrared light.

The photoperiodic stimulus is perceived by the leaves and this can be shown easily by leaving one leaf of a plant exposed while the rest of the plant is covered up. The stimulus must be transmitted in some way to the buds, which then develop flowers. The suggestion is that phytochrome causes a hormone to be produced which affects the buds.

Watch out!

Photoperiodism is a difficult concept so ensure you have a good understanding of the topic.

Checkpoint 6

Why is it now considered unfortunate that historically plants are categorized as short-day or long-day?

Checkpoint 7

Define photoperiod.

Checkpoint 8

Using the adjacent diagram state in each case whether flowering takes place in short-day or long-day plants.

Checkpoint 9

Which part of the plant contains phytochrome and so detects the light?

Exam practice   answers: page 123

1 What is photoperiodism?

2 How does the photoperiod affect flowering?

Nerves

Unlike plants, coordination in animals is brought about by means of two separate but interconnected systems, the nervous system and the endocrine system. The former gives a rapid and short-term response while the latter gives a slower but more lasting one. The cells of the nervous system transmit nerve impulses and are called neurones. You will need to consider their structure and how they are able to transmit information at a remarkable speed in mammals.

Links

Because of their close association with particular organ systems, the activities of certain endocrine glands are dealt with earlier in this section.

Structure of neurones

In the nervous system there are three types of **neurones**.

→ *Sensory or afferent*: bring impulses from the sense organs or receptors into the central nervous system.
→ *Motor or efferent*: take impulses from the central nervous system to the effector organs (muscles or glands).
→ *Intermediate or relay neurones*: receive impulses from sensory neurones or other intermediate neurones and relay them to motor neurones or other intermediate neurones.

Each cell consists of a cell body containing a nucleus and granular cytoplasm containing many ribosomes. These ribosomes are grouped together forming Nissl granules, which are concerned with the formation of neurotransmitter substances. From each nerve cell body arise fine branching structures, **dendrites**. These receive impulses from other nerve cells. Some neurones also have a long membrane-covered cytoplasmic extension, the **axon**, which transmits impulses from the cell body. At its end, an axon divides into branches which form synapses with other neurones.

Peripheral neurones are surrounded by and supported by Schwann cells. In some cases, these grow around the axons of the nerve cells to form a multilayered fatty myelin sheath, found only in vertebrate nervous systems. This acts as an electrical insulator and speeds up the transmission of impulses. The myelin sheath has thin areas at intervals, nodes of Ranvier, which are important in impulse transmission.

Transmission of nerve impulses

→ The **resting potential** is the potential difference between the inside and the outside of a membrane when a nerve impulse is not being conducted.
→ Resting potentials are typically minus values, the minus indicating the inside is negative with respect to the outside. The membrane is said to be polarized.
→ The sodium–potassium exchange pumps (these are transmembrane proteins) maintain the concentration and an uneven distribution of sodium ions and potassium ions across the membrane.
→ The composition of the cytoplasm of an axon is very different from that of the surrounding fluid. Potassium (K^+) and organic anions (COO^-) are higher inside the neurone while the concentration of

Checkpoint 1

In table form list the differences between hormonal and nervous communication using the following headings: origin of stimulus, nature of stimulus, means of transmission, destination of stimulus, receptor, speed of transmission, effects, duration.

Checkpoint 2

Describe the direction of the impulse through a neurone.

Checkpoint 3

Why do myelinated axons conduct impulses faster than non-myelinated fibres of the same diameter?

Checkpoint 4

Is the net charge inside the membrane of a neurone positive or negative?

Checkpoint 5

List in the correct order the sequence of changes in the membrane of an axon between the resting potential condition, the passage of an impulse and the re-establishment of a resting potential.

sodium (Na^+) is higher outside. The membrane is more permeable to K^+ than any of the others.

→ Since the concentration of K^+ ions is higher than outside, they diffuse out. This outward movement of positive ions means that the inside becomes slightly negative.

→ Nerve impulses are due to changes in the permeability of nerve cell membrane to K^+ and Na^+ that leads to changes in the potential difference across the membrane and the formation of action potential.

→ Suitable stimulation of an axon results in change of potential across the membrane from a negative inside value of about −70 mV to a positive inside value of +40 mV. This change is called an **action potential**. The membrane is said to be **depolarized**.

→ The action potential is the result of a sudden increase in the permeability of the membrane to Na^+. This allows a sudden influx of Na^+ which depolarizes the membrane.

→ A fraction of a second after this depolarization the K^+ ions diffuse out and repolarize the membrane. There is an overshoot of K^+ leaving as the K^+/Na^+ pump restores the ionic balance. This is called the **refractory period** during which another action potential cannot be generated, so ensuring a unidirectional impulse and limiting frequency.

The process may be summarized as follows. The action potential causes a small electric current across the membrane and as a portion of the membrane is depolarized, depolarization of the next portion is initiated. There is a series of local currents propagated along the axon. The sodium pump is active all the time and behind the transmission; this pump restores the resting potential. Once the resting potential is restored, another impulse can be transmitted. As the impulse progresses the outflux of K^+ causes the neurone to be repolarized behind the impulse. There is an overshoot of K^+ leaving as the K^+/Na^+ pump restores the ionic balance. During the period between depolarization and repolarization impulses cannot pass. This is known as the refractory period.

When axons have a myelin sheath there is no contact between the axon membrane and the tissue fluid except at the nodes of Ranvier. The result is that the impulse jumps from one node to the next, speeding the overall passage along the axon. Also, the larger the diameter of the axon, the greater the velocity of transmission (see diagram below).

myelin sheath   local circuit   node of Ranvier

direction of impulse conduction

Action point

Draw a labelled diagram of a generalized neurone and show the direction of the nerve impulse.

Grade booster

Research nerve toxins; the risks and benefits of botox treatment.

Checkpoint 6

What is the source of energy used to establish a resting potential?

Checkpoint 7

Name two features of mammalian neurones that increase the speed of transmission of impulses.

Checkpoint 8

What is the importance of the refractory period?

Checkpoint 9

An action potential is an 'all or nothing' response. What does this mean?

Exam practice   answers: page 123

1 Give an account of the general structure of the mammalian motor neurone and describe the way in which the impulse is transmitted along the neurone.

Synapses

Grade booster

Research the action of drugs used as medical inhibitors.

Neurones are not in direct contact with each other but are separated by tiny gaps known as synapses. The main role of the synapse is to convey action potentials between neurones. A range of drugs functions by interfering with the neurotransmitters involved in synaptic transmission.

Structure of a synapse

Most junctions between neurones take the form of chemical synapses. Branches of axons lie close to dendrites of other neurones but do not touch; there is a gap of about 20 μm between them. When impulses are transmitted, this gap is crossed by the secretion of a neurotransmitter from the axon membrane (presynaptic membrane), which diffuses across the space to stimulate the dendritic membrane (postsynaptic membrane). The structure of a synapse is shown below.

Checkpoint 1

What general name is given to chemicals which transmit messages across a synapse?

Synaptic transmission

The arrival of the impulses at the synaptic knob alters its permeability allowing calcium ions to enter. The influx of calcium ions causes the synaptic vesicles to fuse with the presynaptic membrane so releasing the transmitter (acetylcholine) into the synaptic cleft. When the transmitter diffuses across the gap (synaptic cleft) it attaches to a receptor site on the postsynaptic membrane, depolarizing it and so initiating an impulse in the next neurone.

Checkpoint 2

How can there be an unlimited supply of acetylcholine?

The postsynaptic membrane contains specific protein receptors with which the transmitter molecules combine. Once combined, protein channels open up in the membrane, allowing sodium ions to diffuse from the cleft into the postsynaptic neurone. Acetylcholine, when released, is quickly destroyed by enzymes in the synaptic cleft, so its effect is limited and the merging of impulses is prevented. If insufficient acetylcholine is released, the postsynaptic membrane will not be stimulated. The enzyme that destroys acetylcholine is called cholinesterase. The resulting choline and ethanoic acid diffuse back across the synaptic cleft for further use.

Checkpoint 3

By what process do ions cross the post-synaptic membrane when a neurotransmitter substance has caused ion-specific channels to be opened?

Checkpoint 4

What is the function of organelles such as mitochondria and the Golgi apparatus in the synaptic knob of neurones?

Functions of synapses

→ Transmit information between neurones.
→ Pass impulses in one direction only.
→ Act as junctions.
→ Filter out low-level stimuli.

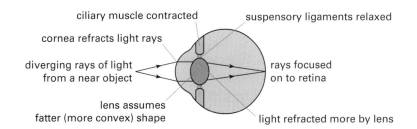

ciliary muscle contracted   suspensory ligaments relaxed

cornea refracts light rays

diverging rays of light
from a near object → rays focused
on to retina

lens assumes
fatter (more convex) shape → light refracted more by lens

The retina

The retina possesses the photoreceptor cells. These are of two types, **rods** and **cones**, and their structure is shown in the diagram below.

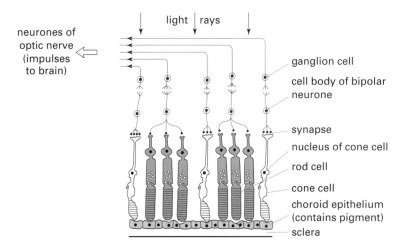

neurones of
optic nerve
(impulses
to brain) ⇐

light rays

ganglion cell

cell body of bipolar
neurone

synapse
nucleus of cone cell
rod cell
cone cell
choroid epithelium
(contains pigment)
sclera

Rod cells are distributed throughout the retina whereas cone cells are concentrated in and around the fovea. Many rod cells share a single neurone and therefore give poor visual acuity (sharpness of vision). Each retinal photoreceptor cell contains a photosensitive pigment. In a rod cell this is rhodopsin or visual purple. Light absorbed by the rod cell causes rhodopsin to split into its constituent parts, opsin and retinal, in a process called bleaching. It is this chemical change that initiates an action potential along the neurone joining the rod cell to the brain.

Colour vision

It is only cone cells that are sensitive to different colours. It is thought that they are of three types, any one being sensitive to blue, green or red light. It is the degree of stimulation of each of these three types which provides the full range of colours visible to organisms with colour vision. This theory is called the trichromatic theory of colour vision.

Test yourself

Describe the condition of the eye when focused on a distant object.

Checkpoint 4

Why do some people hold a book at arm's length in order to read? Why does a bright light help?

Checkpoint 5

Make up a table to compare rods and cones using the headings: relative numbers, distribution in retina, visual acuity, light sensitivity, ability to discriminate light of different wavelengths.

Checkpoint 6

Which part of the retina contains only cone cells?

Checkpoint 7

The diagram of the rod cells shows there are numerous mitochondria present. Suggest how the resynthesis of rhodopsin occurs.

Checkpoint 8

Suggest how we can see other colours besides blue, green or red.

Exam practice   answers: page 125

1 From your knowledge of the retina explain why two small objects close together can be more easily distinguished by rods than cones.

Muscles

Movements of the body are brought about by the contraction of voluntary muscles, which are attached to movable bones of the skeleton. You should look at the structure of skeletal muscle as seen with the light and electron microscopes. To understand how a muscle contracts you need to study the sliding-filament theory.

The sliding-filament theory of muscular contraction

This theory suggests that the muscle filaments do not shorten when the muscle contracts but that they slide between one another. The theory is based on the study of electron micrographs of muscles fixed at different degrees of tension.

Striated or skeletal muscle consists of fibres, each of which is a long multinucleate cell. The cell wall, or **sarcolemma**, encloses several nuclei, cytoplasm and numerous mitochondria. It also contains the contractile elements, which are long protein fibres, the myofibrils. Each myofibril is divided into sarcomeres by cross partitions, the Z lines (see diagram below).

In each sarcomere there are two types of protein fibre:

→ thin **actin** strands project into the sarcomere from the Z lines
→ thicker **myosin** strands between them, which are fixed in position by molecular bonds (see diagram below).

During contraction the actin and myosin filaments slide together

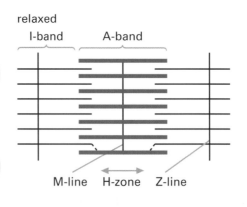

thin filaments (actin)
thick filaments (myosin)

relaxed
I-band   A-band

contracted
sarcomere

M-line   H-zone   Z-line

H-zone, sarcomere and I-band all shorten, the A-band does not change in length

The jargon

A *myofibril* consists of a number of repeated sarcomeres.

Checkpoint 1

Is the A band made up of both actin and myosin or only one of them?

Checkpoint 2

What makes up the I band?

Checkpoint 3

Why does a muscle fibre contain light and dark bands?

Action point

By means of a bullet list describe the changes that take place when a muscle contracts.

Check the net

You'll find information about the role of calcium ions in muscle contraction at shef.ac.uk/uni/projects/mc/excont.html to form a shorter sarcomere (see diagram on previous page). The filaments slide together because of the formation of cross bridges that alternately build and break during contraction.

The ratchet mechanism

The 'ratchet mechanism' may be used to explain how the actin and myosin filaments slide past one another.

→ There are cross bridges between the actin and myosin filaments.
→ When an impulse reaches the neuromuscular junction, acetylcholine is released, which causes the sarcolemma of the muscle to be depolarized. This causes the release of calcium ions, which activate the protein **troponin**. This displaces another protein, **tropomyosin** (which has been blocking the actin filament), away from attachment sites on the actin molecule so myosin and actin can be linked by cross bridges (see diagram below).
→ The bulbous heads along the myosin filaments form these bridges.
→ Each myosin filament has a number of bulbous heads.
→ The clubbed heads of the myosin molecule swing back and forth with the breaking and reforming of bonds.
→ The bulbous heads progressively move the actin filament along, as it becomes attached and detached.
→ The process receives energy by the hydrolysis of ATP.
→ Phosphocreatine is used to regenerate ATP.

The jargon

The *neuromuscular junction* is the point where the effector nerve meets a skeletal muscle.

Action point

Describe the changes that occur in the appearance of a myofibril on contraction.

Checkpoint 4

Using the adjacent diagram how would you describe the proteins tropomyosin and troponin, fibrous or globular?

Examiner's secrets

The sequence of events for the stimulation of a muscle is much the same as the mechanism of synaptic transmission.

Exam practice   answers: page 125

1   Give an illustrated account of the sliding-filament theory of muscle contraction.

Answers
Regulation and control

Homeostasis

Checkpoints

1 Dehydration.
2 Hypothalamus.
3 Superficial arterioles dilate allowing more blood to run through the surface capillaries increasing heat loss through radiation.
4 It switches on the heat loss mechanisms when the temperature of the blood is higher than normal. The heat gain centre is inhibited by the heat loss centre but becomes active when receptors in the skin signal that the environment is getting cooler. Heat production mechanisms and conservation mechanisms are also switched on.
5 They monitor external temperature changes.

Exam practice

1 (a) Hypothalamus.
  (b) (i) Nerve impulses.
      (ii) Vasodilation of arterioles brings more blood through the surface capillaries, with the loss of heat by radiation.
2 (a) Negative feedback means that an increase in some factor causes changes to occur which operate to bring about a decrease in the original factor. This provides a self-regulatory mechanism for maintaining a constant level in the factor. This is the basis of homeostasis and it is important in living organisms that all internal processes fluctuate only marginally from the norm.
  (b) The factor is the level of glucose in the blood.
    • Any fluctuation in glucose concentration initially affects the level of insulin secretion by the β cells of the islets of Langerhans in the pancreas.
    • Increased glucose increases insulin secretion.
    • Increase in insulin increases the uptake of glucose by muscle cells.
    • This promotes oxidation of glucose in the liver or muscle.
    • This promotes the conversion to glycogen by liver cells and inhibits glycogen breakdown.
    • Decrease in insulin reduces uptake of glucose by cells.
    • Lowered blood glucose levels stimulate islet α cells to release glucagon.
    • Glucagon stimulates the breakdown of glycogen.

The kidney

Checkpoints

1 Cortex.
2 Medulla.
3 Ureter.

Exam practice

1 (a) Glomerulus.
  (b) Pressure at X would increase.
      Rate of flow of blood decreases.
      Rate of ultrafiltration would increase or more fluid passes to Bowman's capsule.

Examiner's secrets

In 1(b) it is not sufficient to write that 'blood pressure increases'. You must be more specific in your answer and state that it increases at a specific point.

Functioning of the kidney

Checkpoints

1 Passive.
2 Glucose, amino acids, vitamins, some hormones, urea, uric acid, creatinine, ions, water.
3 Cells, platelets, plasma proteins which are too large to pass through the filter.
4 Longer.
5 The blood gains ions during the descending journey and loses them again from the ascending loop.
6 Base of loop.
7 It increases.
8 Detector – hypothalamus; coordinator – posterior lobe of pituitary (ADH); effector – collecting ducts of kidney.
9 Fall in water potential of the blood.

Exam practice

1 (a)
  • There is a high pressure in the glomerular capillary.
  • The filter is provided by pores in the basement membrane of Bowman's capsule.
  • Particles of a certain size are allowed through, e.g. glucose, etc.
  • All the glucose is reabsorbed in the first convoluted tubule by diffusion and/or active transport.
  (b)
  • The loop of Henle enables the counter-current multiplier to operate.
  • Na^+ ions are actively pumped out of the ascending limb into the medulla, creating a low water potential there.
  • Na^+ diffuses into the descending limb.
  • The walls of the ascending limb are relatively impermeable to water.
  • Therefore reabsorption of water occurs from the permeable descending limb.
  • Water leaves the filtrate by osmosis.
  • The contents of the descending limb become more concentrated.
  • The maximum concentration occurs at the tip of the loop of Henle.

Plant growth substances

Checkpoints

1 There is a reduction in the numbers of fruits and seeds formed when environmental conditions are unfavourable.
2 Broad-leaved weeds have a large surface area and so receive a large dosage of weedkiller when applied; the narrow-leaved grasses do not. There may also be a difference in sensitivity of the different types of leaves.
3 The loss of soil nutrients to broad-leaved weeds and competition for light is reduced.
4 Some are more effective than naturally occurring substances, are easily manufactured, are stable in environmental conditions favourable for growth and can be more selective in the range of species affected.
5 The removal of the apical bud would allow the development of lateral buds. The addition of IAA would reinstate the inhibition.
6 Fruits produce ethene as they ripen. Ethene is a gas that diffuses through the air and causes ripening of unripe fruit.

Exam practice

1 Include all the following points in your answer:
   • Auxins stimulate adventitious root growth.
   • Auxins induce fruit development in the absence of pollination.
   • Auxins can disrupt normal growth pattern making them useful as herbicides.
   • Cytokinins mixed with auxins control differentiation of plants produced from tissue cultures.
   • Anti-ageing properties of cytokinins used to keep cut flowers fresh.
   • Cytokinins prolong the shelf life of fruit or vegetables.
   • Gibberelins influence fruit set.
   • Gibberelins break seed dormancy.
   • Gibberelins break bud dormancy in the spring.

Effect of light on plant growth

Checkpoints

1 In continuous light, the coleoptile splits open and the leaves emerge.
2 Curvature of the shoot towards the light puts the leaves in a good position for photosynthesis.
3 None.
4 To the left.
5 Natural sunlight contains more red light than far-red light.
6 It is the length of the dark period that is crucial.
7 Length of day.
8 A: dark period is short enough for long-day plants to flower but not long enough for short-day ones to do so.
   B: as there is no single long light period and yet long-day plants flower it must be the short dark period which is the critical factor inducing flowering.
   C: the dark period is long enough for short-day plants to flower but not short enough for the long-day ones to do so.
   D: the short light period in the middle of the dark period prevents short-day plants flowering as there is no single long dark period. Each dark period is, however, short enough to induce flowering in long-day plants.
9 Leaves.

Exam practice

1 See definition in text.
2 Reference to long-day, short-day and day-neutral plants, together with an explanation of the link with phytochrome and how flowering is brought about in both long-day and short-day plants is required. Give named examples where possible.

Nerves

Checkpoints

1
Hormone	Nervous
gland	sense receptor
hormone	nervous impulse
bloodstream	nerve fibre
all over body	to a specific point
target organ	effector
usually slow	rapid
widespread	localized
long-lasting	usually brief

2 In at dendrites, out through axon.
3 Action potentials jump from node to node at high speed.
4 Negative.
5 • Ion channels in the membrane open, Na^+ ion channels open first, Na^+ ions flow in generating an action potential.
  • Na^+ ion channels close again and K^+ ion channels open, K^+ ions flow out and the resting potential starts to be re-formed.
  • Na^+ ions pumped out as K^+ ions pumped in; K^+ ions diffuse out rapidly but Na^+ ions diffuse back in only slowly creating the resting potential.
6 ATP.
7 Nodes of Ranvier and diameter of the axon.
8 The impulse can only move in a forward direction.
9 Provided the stimulus exceeds a certain threshold value an action potential results.

Exam practice

1 Your essay should include a well-labelled diagram and eight of the following points in a logical sequence.
  • Diagram of motor neurone with cell body containing nucleus and with short dendrites and a long axon.
  • Lipid myelin sheath surrounding axon.
  • Interrupted at intervals by the nodes.
  • At rest the outside of the axon membrane has a higher concentration of Na^+ ions than the inside.
  • The reverse is true of K^+ ions.
  • This results in a net positive charge of +70 mV on the outside.
  • Stimulation causes an increase in membrane permeability.
  • The arrival of the action potential opens Na^+ channels in the membrane protein.
  • Na^+ ions rush in along the concentration gradient.
  • The polarity reverses and the outside becomes relatively negative.
  • The positive internal charge forces potassium out along its concentration gradient restoring a positive polarity to the outside.
  • Sodium/potassium pump then re-establishes system ready for next impulse.
  • Impulse travels as a wave of depolarization.
  • It may jump from node to node of the myelin sheath.

Synapses

Checkpoints

1 Neurotransmitters.
2 Acetylcholine is broken down by cholinesterase and the components re-form on the other side of the synaptic cleft.
3 Facilitated diffusion.
4 ATP production and the generation of vesicles of transmitter substance.
5 The insecticide inhibits the enzyme which breaks down the neurotransmitter so new action potentials are propagated indefinitely and impulses merge.
6 Excitatory.

Exam practice

1 (a) (i) A, mitochondrion; B, synaptic vesicle.
     (ii) Acetylcholine / neurotransmitter.
  (b) The ions cause the synaptic vesicle to fuse with the presynaptic membrane and the contents are shed into the synaptic cleft.
  (c) (i) Intrinsic proteins.
     (ii) The neurotransmitter combines with the protein (or the receptor site) causing the protein to change shape, thus opening the channels.
     (iii) The influx of sodium ions depolarizes the postsynaptic nerve.
  (d) The mitochondria provide large amounts of ATP which is used to pump the calcium ions out again or to resynthesize neurotransmitter after its breakdown.

The central nervous system

Checkpoints

1 Hypothalamus.
2 Cerebrum.
3 Contains more cell bodies.
4 An automatic response which follows a sensory stimulus.
5 A pathway of neurones involved in a reflex action.
6 The reaction is routine and predictable and does not need analysis and would be wasteful of the brain's capacity.

Exam practice

1 • Cerebral hemispheres and their position.
   • Correct location for intelligence, learning, language, speech, etc.
   • As well as sensory interpretation.
   • Motor areas controlling voluntary movement.
   • Hypothalamus on the floor of the brain and its role in temperature regulation and osmoregulation and regulation of pituitary secretion.
   • Correct position of cerebellum on lower rear area and its function of coordinating subconscious movement for balance, etc.
   • Medulla in the brain stem as a relay centre for nerve tracts and its role in heartbeat, breathing, etc.
   • Accurate mention of meninges and their function.

Examiner's secrets

A plan is essential for this type of essay and a diagram would be valuable. The only way to memorize a diagram is to close your book and attempt to draw from memory until you get it right! The brain is in three parts: fore, mid and hindbrain. Now place the correct labelling against the structures in each region. Then list the functions of each part of each region.

Sensory receptors

Checkpoints

1 Cornea and lens.
2 Retina.
3 Contains blood capillaries to provide food and oxygen.
4 With ageing the lens hardens and the ability to accommodate is lost, making it hard to focus on near objects. The reduction in pupil size (which occurs as light intensity increases) improves the depth of focus.
5

Rods	Cones
numerous	less numerous
around periphery	in centre
one bipolar neurone	each served by own neurone
sensitive to low light	only stimulated by bright light
stimulated by most wavelengths	three types, each selectively responsive to different wavelengths

6 Fovea.
7 ATP resynthesis.
8 Combined stimulation of cones.

Exam practice

1 This question is about visual acuity. Groups of rod cells share a single nervous pathway to the brain so that the stimulation of more than one of a group still only produces a single impulse.

Muscles

Checkpoints

1 Both.
2 Actin only.
3 Where actin and myosin overlap the band appears dark.
4 Tropomyosin – fibrous; troponin – globular.

Exam practice

1 Your answer should cover 10 of the following points in a logical sequence:
   • Banded into Z, A, I and H with a description or labelled diagram.
   • Filaments are thick myosin and thin actin.
   • H, myosin alone; I, actin alone; A, length of myosin.
   • Diagram showing shortening of sarcomere with more overlap.
   • A correct description of the changes in band width due to actin sliding over myosin, with no change in length of either.
   • When the muscle is stimulated calcium ions are released.
   • The protein troponin is activated with the calcium ions binding.
   • This pushes another protein, tropomyosin, away from the attachment sites on the actin molecules.
   • Myosin and actin can now be linked by crossbridges.
   • Show the clubbed head of the myosin molecule, which swings back and forth with the breaking and re-forming of bonds to produce a ratchet mechanism.
   • This is energized by the hydrolysis of ATP.

Examiner's secrets

You must be very precise in your answer and plan the essay very carefully. It must include illustrations which are credited in the mark allocation. Split the essay into two main headings:
   • the sliding-filament theory and include diagrams of the various lines and a diagram showing the actin and myosin filaments in a relaxed state and in a contracted state
   • the ratchet mechanism, which is very difficult to describe and much easier to illustrate using an annotated diagram

Revision checklist
Regulation and control

1	Define the terms homeostasis and negative feedback.	Confident	Not confident **Revise** page 102
2	Explain how blood glucose concentration is regulated with reference to insulin, glucagon and the liver.	Confident	Not confident **Revise** pages 102, 103
3	Describe and explain the production of urine with reference to the processes of ultrafiltration and selective reabsorption.	Confident	Not confident **Revise** page 106
4	Outline the role of plant hormones in growth.	Confident	Not confident **Revise** page 108
5	Explain the role of phytochrome in flowering.	Confident	Not confident **Revise** page 110
6	Describe and explain how an action potential is transmitted in a myelinated neurone.	Confident	Not confident **Revise** page 112
7	Describe with the aid of diagrams the structure and functioning of a cholinergic synapse.	Confident	Not confident **Revise** page 114
8	Draw a diagram of the brain and outline the functions of each region.	Confident	Not confident **Revise** page 116
9	With the aid of a diagram indicate the neurones involved in a reflex action, using arrows to show the direction of impulse transmission	Confident	Not confident **Revise** page 117
10	In the eye, describe the differences in sensitivity and visual acuity as explained by differences in the distribution of rods and cones.	Confident	Not confident **Revise** page 119
11	Explain with the aid of diagrams the sliding filament model of muscle contraction.	Confident	Not confident **Revise** page 120

Microbiology

Recent developments in biotechnology have resulted in major advances in many different areas of medicine. Humans have studied microorganisms and manipulated their growth requirements in vessels called fermenters. Hormones, antibiotics and therapeutic proteins, such as monoclonal antibodies, are some of the cell products that have been mass produced and have been of benefit in the field of medicine. You also need to study the various control and prevention methods used against selected pathogenic organisms in addition to the natural defence mechanisms of the body. You need to appreciate the problems associated with the overuse of antibiotics.

Exam themes

- The growth requirements of microorganisms
- Measurement of microbial population growth
- The general structural features of a fermenter used for large-scale production
- The advantages and disadvantages of batch and continuous culture
- The use of monoclonal antibodies in medicine
- The causes and means of transmission of named diseases
- The social, economic and biological factors in the prevention and control of named diseases
- The role of antibiotics in the treatment of infectious disease
- Antibiotic resistance
- The origin, maturation and mode of action of phagocytes and lymphocytes
- The immune response
- The actions of B lymphocytes and T lymphocytes in fighting infection
- The distinction between different types of immunity
- How vaccination can control disease
- Antibiotics and the development of resistance

Topic checklist

	Edexcel		AQA		OCR		WJEC	
	AS	A2	AS	A2	AS	A2	AS	A2
Types of microbes		●	O		O			●
Growth and culture techniques			O		O			●
Fermentation			O			●		●
Pathogens and human disease		●	O		O			
Protection against disease		●	O		O			
Applications and contemporary issues 1		●	O		O			●
Applications and contemporary issues 2		●	O		O			●
Applications and contemporary issues 3		●	O		O			●

Types of microbes

It is important to study a range of microorganisms in order to appreciate their diversity with respect to size, complexity and reproduction. Therefore you need to consider bacteria, fungi and viruses and describe their general characteristics.

Bacteria

Bacteria are the smallest cellular organisms and are prokaryotes. The kingdom includes both heterotrophic and autotrophic forms. The diagram below shows the structure of a typical bacterial cell.

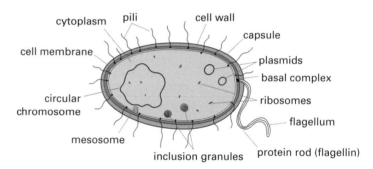

There are several ways of classifying bacteria. One way is by their shape. Bacteria may be:

→ rod-shaped or **bacilli**
→ spherical or **cocci**
→ corkscrew-shaped or **spirilla**
→ comma-shaped or **vibrio**

Further differentiation is often possible according to the way bacteria tend to be grouped, e.g. in pairs, chains or clusters. The filamentous bacteria are of particular interest because most are photosynthetic. This group, formerly called blue-green algae, is now known as cyanobacteria.

Another method of classification is based on the variation in nature of bacterial cell walls. The Gram stain technique distinguishes Gram-negative bacteria and Gram-positive bacteria (see page 140).

Bacteria reproduce asexually by binary fission, although certain bacteria can also reproduce sexually.

Although many bacteria are pathogens and cause deterioration to stored food, it is important to remember that many are beneficial to humans.

Fungi

Although originally classed as plants, fungi are now recognized as a separate kingdom. This is based on the presence of the polysaccharide chitin, found in their cell walls, rather than the cellulose present in plant cell walls. Fungi also lack chlorophyll and are therefore unable to photosynthesize. Fungi are eukaryotic heterotrophs, generally feeding as saprobionts or parasites. The basic structural unit of a fungus is a thread-

Links

Prokaryotes and eukaryotes are discussed on page 24.

Checkpoint 1

List the characteristics of a prokaryote such as a bacterium.

Examiner's secrets

The examiner will be impressed if you can refer to named microorganisms.

Checkpoint 2

What shape is a *Streptococcus* bacterium?

Check the net

You'll find more information on 'The microbial world' at helios.bto.ed.ac.uk/bto/microbes/ microbes.html

Links

Different uses of bacteria can be found in various topics, such as the nitrogen cycle (page 74), and food production and manufacturing processes later in this section.

like multinucleate hypha. A collection of hyphae is called a mycelium. While some fungi cause damage and disease, particularly to plants, many are beneficial to humans, e.g. helping to maintain soil fertility by mineral recycling and decomposing organic matter, food and drink production and the production of antibiotics.

Checkpoint 3

What do you consider to be the most important difference between fungi and plants?

Viruses

Viruses are extremely small and cannot be seen using the light microscope. They enter living cells and multiply with the assistance of the host cells. Viruses cause a variety of infectious diseases. Each virus particle, or virion, is made up of a core of nucleic acid surrounded by a protein coat, the capsid. Most viruses are found in animal cells and those attacking bacteria (bacteriophages) have the nucleic acid DNA. Other animal and plant viruses contain RNA. A widely studied virus is T2 phage, a bacteriophage, which infects *Escherichia coli*.

Links

Antibiotics are discussed on pages 140 and 142.

When a phage attacks a bacterium the capsid attaches to the surface of a host cell and the viral nucleic acid then enters. However, parasitism by phages is not typical of all viruses. In many cases the entire virus penetrates the host cell membrane and its nucleic acid passes to the cell nucleus.

Retroviruses, including HIV, contain RNA. Once inside the host, the RNA serves as a template for the formation of a single-stranded DNA. This molecule in turn acts as a template for the formation of a double-stranded DNA, which attaches to a host cell chromosome. Here, among other functions, it produces viral RNA for retroviruses of the next generation.

Checkpoint 4

For the replication of a virus in its host, why is it not essential for the protein coat of the virus to enter the host cell?

Action point

Draw a T2 phage and a tobacco mosaic virus.

Exam practice   answers: page 144

1 (a) Complete and label the outline diagram below to show three features which may be found in bacteria but not in other groups of microorganisms.

   (b) State two structural characteristics of fungi which bacteria do not possess.

2 (a) Bacteria may be grouped or classified according to the shape of their cells. Complete the table by drawing and naming two of the groups.

Shape of cells	Name of group

   (b) Name the staining technique that may also be used to classify bacteria into two groups, positive and negative.

The jargon

HIV, human immunodeficiency virus, causes the condition known as AIDS.

Links

Genetic engineering is discussed on pages 30, 31 and 182, 183.

Growth and culture techniques

Microorganisms quickly reproduce given a suitable environment. In the laboratory they can be grown on a wide variety of substrates providing they are supplied with suitable physical conditions, nutrients and water. A variety of methods can be used to measure growth rate.

Conditions necessary for growth

Microorganisms require the following conditions for growth.

→ *Nutrients*: in the laboratory nutrients are supplied in nutrient media and include carbon (usually organic such as glucose), nitrogen (organic and inorganic form) and growth factors such as vitamins and mineral salts.
→ *Temperature*: the range of 25–45 °C is favourable for the majority of bacteria as all growth is regulated by enzymes.
→ *pH*: most bacteria are favoured by slightly alkaline conditions (pH 7.4) whereas fungi prefer neutral to slightly acid conditions.
→ *Oxygen*: many microorganisms are aerobic and are termed 'obligate aerobes'. Some, while growing better in the presence of oxygen, can nevertheless survive in its absence; these are called 'facultative anaerobes'. Others cannot grow in the presence of oxygen and are called 'obligate anaerobes'.
→ *Water*: required for metabolic reactions.
→ *Osmotic factors*: to ensure the absorption of water, the water potential of the external environment must be higher (less negative) than the cell contents. Thus most microbes cannot grow in environments with a high solute concentration.

Culture techniques

The growth of a culture of individual cells follows the same pattern as the growth of any population, i.e. lag phase, exponential phase, stationary phase and death phase.

Bacteria and fungi are cultured (grown) on, or in, media that are designed to supply the cell with all their nutritional requirements. Aseptic techniques, in which the apparatus and equipment are kept free of microorganisms, are used to prevent contamination of bacterial cultures and the surrounding environment (see diagram on next page).

Methods of measuring growth

There are several different methods of measuring growth.

→ Rough estimates of growth rates can be made by regularly measuring the diameter of a bacterial or fungal colony as it spreads from a central point to cover the surface of a solid growth medium.
→ Unit samples from broth cultures can be filtered, dried and weighed.
→ Changes in cell density can be measured by the following methods.
  → *Haemocytometer* measures total counts (both living and dead cells).

Links

Enzymes and temperature are discussed on page 20.

Action point

Explain, in biological terms, how most food preservation methods use the principle of removing conditions necessary for growth.

Checkpoint 1

What changes do you think may occur in cells adapting to a new culture medium?

Links

The sigmoid growth curve for population is looked at on page 76.

Examiner's secrets

This section of the topic consists almost entirely of practical work. An examiner will be impressed if you can incorporate your practical experience into your exam answers and be aware of the safety issues and potential hazards associated with microbiology.

→ *Serial dilution* and plating on to a solid nutrient medium (see lower diagram, this page) allows viable counts (living cells only) to be made by counting the number of colonies. Here it is assumed that one cell divides to form a colony.

  → *Photometry*: a colorimeter can be used to measure light intensity and turbidity (cloudiness) as cell numbers increase.

Checkpoint 2

What advantage may be gained from choosing to plot the log of cell number counted in a culture over a period of time?

Checkpoint 3

By what process does one cell give rise to a colony?

Checkpoint 4

What method would a bacteriologist use to estimate the changes in the numbers of viable cells in a culture over a period of time?

equipment, e.g. inoculating wires, petri dishes

sterilization in a pressure cooker or autoclave to kill stray microorganisms

flame sterilization of the inoculating wire before and after transfer of bacteria

media, e.g. nutrient agar, nutrient broth. Complete culture requirements in liquid form (broth) or solidified with special jelly (agar)

source of bacteria, e.g. agar slopes of pure cultures

pouring of plates or slopes

inoculation

incubation at a constant temperature, e.g. 25°C

• **counting** of colonies on plates
• **observation** of colonies
• **staining** of cells

disposal of plates; they are first sterilized at 121°C for 15 minutes

Checkpoint 5

Why is it necessary to sterilize at a temperature of 121 °C for 15 min?

transfer
1 cm³   1 cm³ 1 cm³ 1 cm³ 1 cm³

original culture

tubes containing 9 cm³ of sterile saline

serial dilutions are carried out so that each colony that grows, when a sample is plated, represents a single cell or spore in the original sample

concentration	1	10^{-1}	10^{-2}	10^{-3}	10^{-4}	10^{-5}
dilution factor	0	10	100	1000	10,000	100,000

Exam practice   answers: page 144

1 Describe and explain the procedure that you would use to safely obtain a pure culture of one species of bacterium from a tube containing a mixed bacterial suspension in nutrient medium.

Fermentation

Biotechnology may be defined as the industrial application of biological processes. For many of these processes the key steps are conducted in a fermenter vessel. You should understand the difference between batch and continuous fermentation and how industrial processes involve the need for aseptic entry of material and the control of conditions to ensure optimum yield of product.

Fermenter design

Checkpoint 1

Define fermentation.

A sterile closed fermenter vessel excludes the possibility of contamination by other organisms (see diagram below).

Checkpoint 2

Why is fermentation a misnomer?

input of nutrients

used air out

temperature probe

turbine (to mix contents)

drive shaft (rotates for mixing)

air (dispersed as bubbles in culture medium)

air filter

air inlet

drive motor

pH probe

oxygen concentration probe

baffle (helps contents to mix)

jacket (to maintain temperature)

seal

drain valve (to extract product)

Checkpoint 3

What are the advantages and disadvantages of single-cell protein production?

Checkpoint 4

What is the purpose of fixed aeration?

The products of fermentation may be:

→ a dissolved substance that accumulates in the medium, e.g. the antibiotic penicillin is produced by the fungus *Penicillium*; in this case the used culture medium, after filtering, is kept and processed
→ the microorganism itself, e.g. single-cell protein (mycoprotein) manufacture; in this case the medium may be recycled and the solid processed

Batch versus continuous cultivation

In batch fermentation, as the name suggests, for a particular reaction to occur, growth is allowed to continue up to a specific point at which the fermenter is emptied and the product extracted. The fermenter has to be cleaned and sterilized in readiness for the next batch.

In continuous fermentation, once the fermenter is set up the used medium and products are continuously removed. The raw materials are added throughout.

Antibiotic production

For the commercial production of penicillin, the fungus *Penicillium notatum* (or *P. chrysogenum*) is grown in a batch culture in a fermenter. The antibiotic is secreted into the culture solution by the fungus after the initial growth phase is over and when glucose is depleted. The diagram below shows the sequence of processes involved in the commercial production of penicillin.

nutrients and fungal culture

penicillin extraction by filtration
solid extracted
(*Penicillium* mycelium)

concentration

liquid extracted

air

fermenter

temperature and pH both controlled

immobilized enzyme reaction (for example, natural penicillin is modified to semi-synthetic forms with greater stability)

filtration, cleaning, drying

bulk semi-synthetic penicillins

Antibiotic production is a secondary metabolism, i.e. at a period in the life of the fungus when there is a change away from its optimum conditions. The primary metabolism is the norm, when the fungus is breaking down glucose to release energy and producing an increase in its own biomass (see graph below).

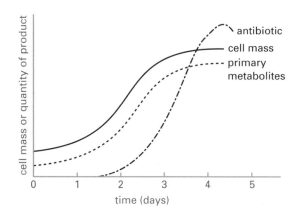

cell mass or quantity of product

antibiotic
cell mass
primary metabolites

time (days)

Checkpoint 5

What are the advantages and disadvantages of continuous fermentation?

Checkpoint 6

What is the importance of routine measurements of pH, temperature and oxygen concentration in a fermenter producing penicillin?

Checkpoint 7

Suggest how *Penicillium* when free living uses penicillin when food sources are depleted.

Links

See also immobilized enzymes on page 21 and antibiotics on pages 140, 141.

Exam practice   answers: page 144

1 (a) With reference to the diagram above explain the importance of each of the following to the industrial manufacture of penicillin.

   (i)   The cooling jacket.
   (ii) The pH probe.
   (iii) The air filter.
   (iv) The inclusion of ammonium as one of the nutrients.

   (b) Suggest why it is often cheaper to use living organisms than conventional laboratory techniques to produce chemicals.

Pathogens and human disease

Another aspect of microbiology is the effect of the different pathogenic organisms and their control. Bacteria, viruses and fungi can be pathogens of humans, their crops and their domestic animals and the consequences can be of great economic significance.

Disease transmission

Many of the important diseases in humans are caused by **pathogenic** bacteria or viruses. These may be transmitted from one person to another through the air, via contaminated food and water, by direct contact or with the aid of vectors. A disease that can be passed from host to host is termed an **infectious** disease. Diseases spread by direct contact are described as **contagious**. Vectors are organisms that carry pathogens from one organism to another, e.g. mosquitoes act as vectors for the protoctistan *Plasmodium*, which causes malaria.

Airborne transmission is also called droplet infection, and is the main way in which diseases of the lungs and respiratory passages are spread. The microorganisms are expelled in tiny droplets of saliva and mucus as a result of sneezing, coughing, or even just breathing. Bacterial diseases spread this way include whooping cough and tuberculosis, while viral infections include influenza and measles.

Food and water can become contaminated with bacteria if sanitation and hygiene are poor. *Salmonella* food poisoning is spread in or from the undercooked meat of infected animals, while diseases such as cholera and typhoid are mainly spread through the contamination of water by the untreated faeces of infected individuals.

Selected diseases, their control methods and treatment

Bacterial diseases

Salmonella food poisoning is caused by a Gram-negative rod-shaped bacterium. Poisonous substances, called **toxins**, produced by the bacteria affect the gut lining causing diarrhoea and vomiting. It is commonly found in animal intestines and may be transferred to meat at animal slaughter. The bacterium multiplies during storage of infected food. Good hygiene practices are essential to avoid food poisoning. These include thorough cooking, storage in cool conditions, separation of cooked and uncooked meat, the prevention of contamination from carriers. Analysis of antigenic types may enable the tracing of the source of an infection.

Cholera is caused by a Gram-negative bacterium, endemic in some areas of the world. Toxins affect the gut lining causing watery diarrhoea leading to severe dehydration and possibly death. Humans can act as a reservoir and as carriers. This disease is often spread by infected water in which the organism does not multiply but may become easily widespread. Protective measures are a clean water supply, the safe disposal of sewage and the hygienic preparation of food. Powerful antibiotic therapy is possible but treatment is largely by rehydration.

Links

Parasites are discussed on page 64.

Examiner's secrets

Check which diseases you are required to study for your specification.

Checkpoint 1

Why is it important to allow a frozen chicken to thaw completely before cooking?

Action point

List the methods that should be used to prevent microbial contamination of foods.

The jargon

An *endemic* disease is always present at low levels in an area.
A *carrier* is infected by a disease organism, shows no symptoms, but can pass the disease on to another individual.
The *reservoir* is the place where a pathogen is normally found.

as the urine seeps up the stick. When the hCG-antibody complex reaches the test region in the stick, they bind with the immobilised antibodies there and are held firmly in position. As the immobilised antibody complex becomes more and more concentrated there is a colour build-up. If the test is positive a coloured band becomes visible through a transparent window.

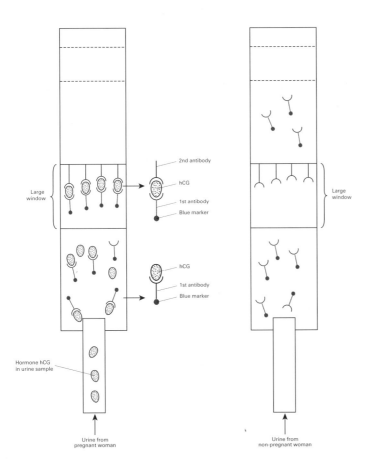

Checkpoint 4

What is the function of the hormone hCG?

Checkpoint 5

Why is urine used to test for pregnancy?

Action point

Compare the reaction in the large window in the two diagrams

Examiner's secrets

Therapeutic means the cure of infections after they have been diagnosed, so limit your answer accordingly.

Exam practice   answers: page 146

1 Describe the production and therapeutic applications of monoclonal antibodies.

Applications and contemporary issues 2

Checkpoint 1

Explain how the use of antibiotics in animal feed leads to the development of resistance in bacteria in humans.

Antibiotics are chemicals produced by micro organisms, which at low concentration inhibit the growth of pathogens. Like disinfectants the chemicals produced interfere with the growth of microorganisms. They may either kill microbes directly – bactericidal, or prevent them from growing – bacteriostatic. To be effective all antibiotics must have selective toxicity, that is, they should kill or inhibit the growth of microorganisms without damaging the host.

Antibiotics

Antibiotics can be divided into two groups:

Checkpoint 2

What is the difference between bacteriostatic and bactericidal?

→ Antibiotics effective against a wide range of pathogenic bacteria are called **broad spectrum** antibiotics, e.g. tetracyclines. These antibiotics affect widespread microbial processes such as protein synthesis.

Checkpoint 3

What is the difference between broad spectrum and narrow spectrum antibiotics?

→ Antibiotics effective against a limited range of pathogens are called **narrow spectrum** antibiotics, e.g. penicillin. These antibiotics are effective against only a few types of bacteria and affect more specific processes such as cell wall formation.

Checkpoint 4

Give an example of a broad spectrum antibiotic.

Bacteria are stained for identification. The Gram stain technique distinguishes two types of bacteria, Gram-positive and Gram-negative. The different staining properties of these two groups of bacteria are due to differences in the chemical composition of their cell walls.

Watch out!

Don't confuse the cell wall structure of Gram-positive and Gram-negative bacteria.

→ **Gram-negative bacteria** have chemically complex walls where a thinner layer of peptidoglycan (murein) is surrounded by an outer layer of lipoprotein and lipopolysaccharide. They do not retain dyes like crystal violet, e.g. *Salmonella* sp. They appear red when viewed under the microscope. The bacteria are resistant to the antibiotic penicillin.

The jargon

Lysozyme causes *lysis*, a dissolving of the walls of Gram-positive bacteria and so they swell by osmosis and burst.

→ **Gram-positive bacteria** lack the lipopolysaccharide in their walls and do retain crystal violet. These are more susceptible to antibiotics and to the enzyme, lysozyme, which occurs in human tears, than Gram-negative bacteria. The Gram-positive bacteria include *Staphylococcus* and *Streptococcus*. This group appears a violet colour when viewed under the microscope.

How do antibiotics work?

Links

Micoorganisms on page 122.

Antibiotics interfere with the growth or metabolism of the pathogen in a variety of ways.

→ Penicillin inhibits the enzymes that are involved in the formation of the cell wall.

Penicillin is an inhibitor of the enzyme, glycoprotein peptidase, which links together the peptidoglycan chains that make up much of the bacterial cell wall (peptidoglycan is similar to glycoprotein). When a newly formed bacterial cell wall is growing it secretes enzymes which make small holes allowing the cell wall to stretch, and new peptidoglycan chains link up across the gaps. Penicillin stops the peptidoglycan chains linking up. However, the enzymes keep making new holes, so the cell wall becomes progressively weaker. As the wall weakens and the cell takes up water and swells, the cell wall is unable to withstand the pressure and the cell bursts. As human cells don't have cell walls, penicillin is harmless to us.

→ Streptomycin inhibits growth by locking on to the ribosomes of the bacteria preventing protein synthesis taking place. Because bacterial ribosomes are different from those of the human, streptomycin does not interfere with protein synthesis in the patient taking the drug.

Grade booster

Research the name of 3 plant-based drugs that are currently in use for the treatment of clinical disease and 3 antibiotics that are produced by microorganisms.

Grade booster

Questions often 'clue' you into the answer. In this question first describe the structure of the cell wall then explain the effect of penicillin on the cell wall formation.

Exam practice   answers: page 146

1 (a) The diagrams show parts of two bacterial cells.

A — Peptidolycan murien / Membrane

B — Peptidolycan murien / Membrane

   (i) Complete the table to show the results of the Gram test on these cells.

Cell	Colour after staining	Positive or negative
A		
B		

   (ii) Name 2 chemical components of the structure labelled X in cell B.

  (b) Bacterial infections, such as septicaemia, are treated with antibiotics. Care needs to be taken to prescribe a suitable antibiotic as the effectiveness of an antibiotic in a given situation depends upon the type of bacterium involved and the mode of action of the antibiotic.

   (i) Describe how penicillin is more effective against Gram-positive rather than Gram-negative bacteria.

   (ii) Suggest why antibiotics that act on protein synthesis are described as 'broad spectrum'.

Applications and contemporary issues 3

Because of the overuse of antibiotics, antibiotic strains of bacteria are continually appearing by mutation. There is a constant battle to find new antibiotics against new resistant strains of bacteria.

Antibiotic resistance

Many bacteria that were previously susceptible to antibiotics have now become resistant. Resistance to a chemical poison is the ability of an organism to survive exposure to a dose of that poison which would normally be lethal to it. This resistance has arisen because of random mutations. Most gene mutations are recessive and rarely expressed in combination with a dominant gene. However, repeated exposure to antibiotics has led to more bacteria surviving and passing on resistant genes.

→ The use of antibiotics to treat disease in the human population has had obvious benefits. However, in agriculture many farm animals are reared indoors so that they grow more rapidly, for example, poultry kept in confined areas. In these crowded conditions there is a greater risk of disease spreading therefore broad-spectrum antibiotics are often added to animal food in an attempt to prevent disease. Although the prophylactic use of small quantities of antibiotics in animal feed has led to healthier, faster-growing animals, this widespread increased use of antibiotics has led to the development of resistance among many species of bacteria.

→ In the past antibiotics have perhaps been prescribed too readily by many doctors in general practice (GPs). Reducing the number of antibiotics in use means that fewer bacteria are exposed to them and reduces the chances of resistant strains appearing.

→ Resistance arises by mutations occurring randomly within populations of organisms which then confers an advantage in the presence of that antibiotic. This may be the ability to produce an enzyme which breaks down the antibiotic. For example, some bacteria have developed an enzyme, penicillinase, which renders penicillin ineffective. In the presence of penicillin non-resistant forms are destroyed. There is a selection pressure favouring the resistant types. The greater the quantity and frequency of penicillin use, the greater the selection pressure.

→ The problem has been made worse by the discovery that resistance can be transmitted between individuals of the same species. There is evidence that the resistance may be passed from one organism to another on plasmids during conjugation (sexual reproduction). This means that a disease-causing organism can become resistant to a given antibiotic even before the antibiotic is used against it.

→ MRSA (methicillin resistant *Staphylococcus aureus*) has developed a resistance to several antibiotics. Antibiotics are widely used in hospitals, especially to prevent infections occurring from surgery.

Checkpoint 1

What is meant by the term resistance?

Checkpoint 2

Why has antibiotic resistance in bacteria increased?

The jargon

Prophylactic means applying antibiotics to prevent infection.

Checkpoint 3

Name a bacterium that has developed resistance to several antibiotics.

Checkpoint 4

Why are viruses unaffected by antibiotics?

This organism originated in Australia and within ten years has spread worldwide. A sample of *Pseudomonas aeruginosa* has been identified as being resistant to all clinical antibiotics. Such organisms are becoming a more common problem and the antibiotics available to treat them are becoming very limited in number.

→ Viruses do not have a cell wall and antibiotics are ineffective against them.

Links

Mutation, natural selection and antibiotics.

Grade booster

This is a very topical issue. Be prepared to answer questions on antibiotic resistance.

Grade booster

Research the extent to which an understanding of the contributory causes of hospital-acquired infections has led to improved codes of practice.

Exam practice   answers: page 147

1   MRSA is a type of *Staphylococcus aureus* of which some strains have become resistant to the class of antibiotics called methicillin used to treat it. The diagram shows the results of an antibiotic sensitivity test on an individual with MRSA, showing the diameters of the inhibition zones in millimetres.

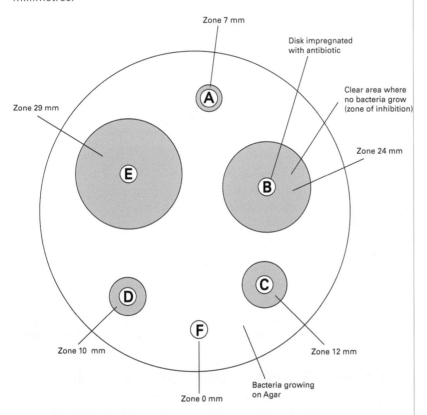

To decide whether the bacterium is resistant to a particular antibiotic the following table of standards is used. The bacterium is said to be resistant if the clear zone of inhibition around the disc is less than the value shown in the table.

Diameter of inhibition zone	A	B	C	D	E	F
Resistant if less than	11	12	9	11	12	14

(a) To which 3 of the antibiotics is the bacterium resistant?

(b) What would you need to know about the antibiotics in order to be able to use these results to diagnose MRSA?

(c) Which of the antibiotics is most likely to be effective against MRSA?

(d) Explain how mutation and natural selection are involved in the development of resistance by certain bacteria to antibiotics.)

Answers
Microbiology

Types of microbes

Checkpoints

1 No distinct nucleus; DNA not incorporated in chromosomes but made up of a single circular strand; no spindle forms at cell division; membrane-bound organelles are absent.
2 Spherical.
3 Chitin cell wall in fungi, cellulose in plants; no chlorophyll in fungi.
4 As only the viral DNA enters the bacterial cell and the protein coat remains on the outside, it is the nucleic acid of the virus that carries the information to produce the protein coat. Inside the host cell fresh 'coat' proteins are made on the host cell's ribosomes.

Exam practice

1 (a) Completed diagram labelled with three of the following: cell wall containing murein/no cellulose or chitin present; circular DNA lying free in the cytoplasm; no nuclear membrane present; mesosomes; flagella; mucilaginous capsule; pili.
  (b) Cell wall contains chitin; consist of hyphae; form a mycelium; no distinct cell boundaries/multinucleate; multicellular; sporangia.
2 (a) Round shape – coccus
     Rod shape – bacillus
  (b) Gram stain

Growth and culture techniques

Checkpoints

1 During the lag phase hydrolytic enzymes specific to the food source are produced.
2 The curve of exponential growth, when presented as a log plot, becomes a straight line, making interpretation easier.
3 Asexual reproduction.
4 Dilution plate method since only live cells give rise to a fresh colony on a dilution plate.
5 Gram-positive bacteria, e.g. *Clostridium*, produce endospores which are resistant to boiling.

Exam practice

1 Your answer should include 10 of the following points in a logical sequence:
   • Aseptic or sterile technique used.
   • Inoculating loop or Pasteur pipette sterilized in the flame of a Bunsen burner or an autoclave or irradiation.
   • Remove cotton wool plug or stopper and flame neck of culture bottle.
   • Streak out with a sterile loop or spread with a sterile spreader or drops on an agar plate.
   • Stress that plates are sterile.
   • Ensure lid of petri dish is exposed to air for a minimum period of time.
   • Seal plates with tape.
   • Incubate at a stated temperature (25 °C or below) for one to two days.
   • Select an isolated colony.
   • Swab work area to maintain aseptic technique.
   • Explanation of asexual reproduction by isolated colony.
   • Stress the importance of a temperature below 25 °C to prevent the development of pathogenic bacteria.
   • Stress the reason for sealing the plate to prevent contamination into the air.
     • The correct disposal of plates using an autoclave.

Examiner's secrets

This question should prove straightforward if you have carried out the procedure in the laboratory. It demonstrates the importance of revising practical work. The question requires you to describe the technique of subculturing. Note how the word 'safely' requires you to mention points such as not leaving petri dishes open but sealing them; inoculating equipment should be sterilized before *and* after use; sterilize work area; safe disposal of plates at end of procedure. A well-annotated diagram could be used to gain many of the marks. Make every effort to show the examiner that you have had experience of practical work.

Fermentation

Checkpoints

1 The biochemical process involving the culture of microorganisms in an aqueous suspension in a culture vessel known as a fermenter.
2 Strictly speaking fermentation means the anaerobic breakdown of glucose to ethanol or lactic acid. Nowadays the word is taken to include aerobic processes.
3 A much faster growth rate is achieved compared with agricultural products, with doubling time measured in hours; a wide range of waste materials, e.g. cheese whey, can be used as substrates. It is a healthy food high in protein, high in fibre and low in cholesterol.
4 For maximum growth of aerobes, also helps to mix the culture to improve contact with nutrients (with the help of a stirrer).
5 Advantages: quicker, no need to empty, clean and refill; by adjusting addition of nutrients the rate of growth can be maintained at a constant level to provide maximum yield. Disadvantages: only suited to the production of biomass or metabolites associated with growth; a sophisticated monitoring technology is required.
6 Monitoring the process of product synthesis; checking that the culture is not contaminated by other organisms causing alternative reactions to occur.
7 To destroy competitors which may be competing for the same food source.

Exam practice

1 (a) (i) *Penicillium* generates heat during its respiration which would denature its own enzymes; the jacket maintains the optimum constant temperature for enzyme activity.

(ii) It adjusts or maintains a constant or optimum pH; if the pH is too high the rate of reaction will slow due to the inactivation of enzymes.

(iii) It removes or prevents the contamination by airborne microorganisms, which would reduce the production of antibiotic due to the introduction of toxins.

(iv) Ammonium is a source of nitrogen for the production of amino acids and proteins by the microorganism.

(b) As microorganisms contain all the necessary enzymes to carry out complex chemical reactions, the processes are carried out at lower temperature with less electrical energy needed and so costs in the long term are reduced. The enzymes are also a renewable source.

Pathogens and human disease

Checkpoints

1 The recommended cooking time would be insufficient to kill bacteria if any part remained frozen.
2 Since the parasites are the target of specific immune attack in the primary host, they take temporary refuge by invading a secondary host.
3 Both invade humans; both invade the blood system; both reinfect more blood cells.
4 It lives in another living organism and causes it harm.

Exam practice

1 Your answer should include a range of microbes with the appropriate control measures.
Protoctista: malarial parasite (*Plasmodium*) or other suitable example.
Fungi: ringworm, athlete's foot, rust, mildew.
Bacteria: tetanus, *Salmonella*, cholera, or any named pathogen.
Virus: HIV, influenza, poliomyelitis, rubella, tobacco mosaic virus, etc.
- Some discussion on prevention being more desirable than the patient actually suffering the disease, stressing the economic advantages of preventative medicine.
- Control of bacterial and viral diseases by vaccination to induce active immunity.
- Vaccination programmes can be very successful, e.g. smallpox, stressing that a high proportion of the population possess immunity.
- The prevention of disease by barriers to infection, e.g. various named measures to exclude mosquitoes including the isolation of sufferers in quarantine.

- Antibiotics to control bacterial disease.
- A discussion of the problem of overuse of antibiotics resulting in resistance.
- Named example of resistance.
- Food hygiene.
- Example of organism causing food poisoning.
- Food inspection.
- Food preservation methods.
- Destruction of vectors, e.g. by insecticide use against the mosquito.
- A problem of insecticide resistance.
- Methods of killing mosquito, e.g. oil, biological control, drainage of swamps.

Protection against disease

Checkpoints

1 Thymus gland.
2 Viruses.
3 T-lymphocyte.
4 Bacteria.
5 Memory cell/B-lymphocyte.
6 No memory cells are formed.

Exam practice

1 Your answer should include the following points and must contain the definitions of antigens and antibodies.
- Antigens are substances which trigger the immune response. They are known as self antigens if they belong to the host and non-self antigens if they belong to a foreign source.
- Antibodies are specific proteins, known as immunoglobulins.
- They are produced by the B-lymphocytes.
- When their receptors detect an antigen the B-cells which carry the correct antibody for the introduced antigen multiply rapidly to produce a clone.
- Some of these become antibody secretors, known as plasma cells.
- Others remain in the circulation as memory cells, but survive in the body ready to clone immediately there is any future invasion by the same antigen.
- A second invasion produces a more rapid response compared with the first.

Applications and contemporary issues 1

Checkpoints

1 A group of identical antibodies that are specific to a single known antigen.
2 Pregnancy testing.
3 Deliver drugs/magic bullets to tumour cells.
4 To stimulate the ovaries to produce hormones needed to maintain a healthy pregnancy.
5 The hormone hCG is excreted in the urine and high levels indicate pregnancy.

Exam practice

1 The production part of the question is detailed in the adjacent diagram. Therapeutic applications should include some detail of the following:
   • Purification of important chemicals, such as interferon, to remove organic contaminants.
   • Combined with cancer treatment to target the surface of cancer cells. This site-specific release of toxins destroys the cancer cells without harming healthy tissue.
   • Prevention of rejection of transplants.

Applications and contemporary issues 2

Checkpoints

1 It is thought that antibiotic residues may accumulate in the food chain when humans eat meat. This overuse of antibiotics has greatly accelerated the spread among bacteria of resistance to antibiotics. The development of resistance occurs as follows. In a large enough population of bacteria a small number are naturally resistant through random genetic variation. Exposure to antibiotic eliminates the susceptible majority. The small number of resistant survivors multiply rapidly in the absence of competition. The population becomes more or less totally antibiotic-resistant.
2 Bacteriostatic chemicals stop the growth of bacteria whereas a bactericidal chemical kills bacteria.
3 Antibiotics effective against a wide range of pathogenic bacteria are called broad spectrum whereas those effective against a limited range are called narrow spectrum.
4 Tetracycline.

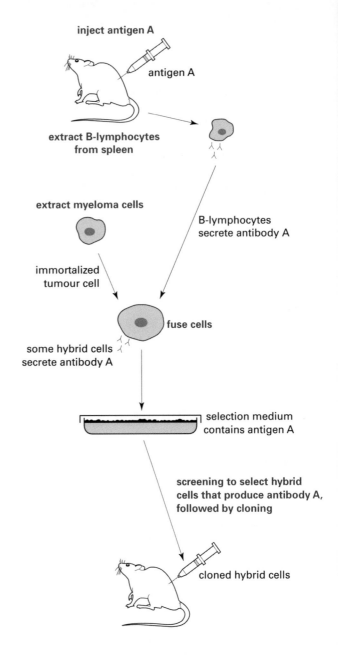

inject antigen A

antigen A

extract B-lymphocytes from spleen

B-lymphocytes secrete antibody A

extract myeloma cells

immortalized tumour cell

fuse cells

some hybrid cells secrete antibody A

selection medium contains antigen A

screening to select hybrid cells that produce antibody A, followed by cloning

cloned hybrid cells

injection into the peritoneal cavity of another mouse

hybridoma (tumour) forms

monoclonal antibody A harvested from peritoneal cavity

Exam practice

1 (a) (i) A – violet – positive

B – pink – negative

(ii) Protein and Lipopolysaccharide.

(b) (i) Penicillin inhibits the formation of cross linkages in the cell wall of Gram-positive bacteria; the murein/peptidoglycan is affected wall is weakened; water is taken up by osmosis; lysis occurs, disrupting cell wall; bacteria are killed; Gram-negative walls are protected by a layer of lipopolysaccharide.

(ii) Broad spectrum – act on a wide range of bacteria and interfere with protein synthesis. This is a feature of all bacteria since it is a common metabolic process.

Applications and contemporary issues 3

Checkpoints

1 Resistance is the ability of an organism to survive exposure to a dose of that poison which would normally be lethal to it.

2 Due to the overuse of antibiotics both to treat patients and in animal feed.

3 MRSA.

4 Viruses do not have a cell wall.

Exam practice

1 (a) A, D and F.

(b) Which are methicillins.

(c) E.

(d) Random mutations cause a change in DNA; causing a different protein to be made; bacteria with advantageous alleles survive; pass on allele to daughter cells.

Revision checklist
Microbiology

By the end of this chapter you should be able to:

1	List the main features of bacteria, fungi and viruses.	Confident	Not confident **Revise** pages 128, 129
2	Describe and explain in the conditions necessary for the growth of bacteria.	Confident	Not confident **Revise** page 130
3	Describe and explain aseptic techniques in the culture of bacteria.	Confident	Not confident **Revise** page 130
4	In a fermenter explain the purpose of a cooling jacket, pH probe, oxygen probe, rotating device and air filter.	Confident	Not confident **Revise** page 132
5	Outline the causes and means of transmissions of malaria AIDS/HIV and other selected diseases.	Confident	Not confident **Revise** page 135
6	Define the terms immune response, antigen and antibody.	Confident	Not confident **Revise** page 136
7	Describe the formation and mode of action of B and T lymphocytes.	Confident	Not confident **Revise** page 136
8	With reference to immunity, distinguish between the terms active and passive, natural and artificial.	Confident	Not confident **Revise** page 137
9	Define the term monoclonal antibodies and list their uses.	Confident	Not confident **Revise** page 138
10	Distinguish between the cell wall structure of Gram-positive and Gram-negative bacteria.	Confident	Not confident **Revise** page 140
11	Distinguish between the mode of action of the antibiotics tetracycline, penicillin and streptomycin.	Confident	Not confident **Revise** pages 140–141
12	Discuss the problem of antibiotic resistance.	Confident	Not confident **Revise** page 142

Reproduction and genetics

In addition to increasing numbers, sexual reproduction produces offspring that are genetically different and so is a means of increasing variety, enabling plants and animals to adapt to changing conditions. Flowering plants have become the dominant plant group on Earth and humans the dominant animal group. Although humans have bred plants and animals for thousands of years, it is only in recent times that they have begun to understand the underlying genetic principles. This has enabled genetic engineers to manipulate human DNA as well as that of animals and crop plants of economic importance. You should be aware of the potential hazards of this course of action.

Exam themes

- Describe and explain human gametogenesis
- The role of hormones in the menstrual cycle, pregnancy, birth and lactation
- The behaviour of chromosomes during meiosis
- How meiosis and fertilization can lead to variation
- How genetic screening is carried out, its advantages and disadvantages and the need for genetic counselling
- The theoretical basis of genetic fingerprinting
- The cloning of plants from tissue culture and the advantages and disadvantages of cloning
- Benefits and concerns of developing GM crops
- Beneficial applications of the Human Genome Project

Topic checklist

	Edexcel		AQA		OCR		WJEC	
	AS	A2	AS	A2	AS	A2	AS	A2
Flower structure and gamete development	○							●
Pollination and fertilization	○							●
Seed development and germination								●
Human reproduction	○		○	●				●
Hormonal control of reproduction	○		○					●
Fertilization and development	○		○	●				●
Meiosis	○	●	○	●		●		●
Variation and evolution	○	●	○	●	○	●		●
Mendel's laws				●		●		●
Deviations from Mendel's laws and sex linkage				●		●		●
Mutations	○	●		●		●		●
Human Genome Project	○	●						●
Genetic counselling and gene therapy	○	●		●		●		●
Genetic fingerprinting		●		●		●		●
Applications of reproduction and genetics	○	●		●	○	●		●
Tissue culture	○			●		●		●
Genetically modified organisms	○			●		●		●

Flower structure and gamete development

The flowering plants or angiosperms are the most successful of all terrestrial plants. The flower is the organ of reproduction and is usually hermaphrodite. In angiosperms the female part, the ovule, is never exposed but is enclosed within a modified leaf, the carpel. You will need to study the structure of a typical flower as well as how it produces the male and female gametes, pollen and ovules.

The jargon

A *hermaphrodite* is an organism with both male and female organs.

Check the net

You'll find some good exercises on flower structure at wsuonline.weber.edu/course.botany. 130/unit1_la.htm

Checkpoint 1

How do flowers attract insects?

Checkpoint 2

What is the nutritional value of nectar to insects?

Checkpoint 3

Describe the adaptations of pollen grains to insect pollination.

Flower structure

A typical flower is made up of four sets of modified leaves, each set being referred to as a whorl.

→ *Calyx* (outer whorl): sepals, which protect the flower in bud.
→ *Corolla*: petals, brightly coloured to attract insects.
→ *Androecium*: stamens, which comprise a long filament at the end of which are the anthers which produce pollen grains containing the male gamete. As well as supporting the anther the filament contains vascular tissue that transports food materials necessary for the formation of pollen grains. The anther is usually made up of four pollen sacs arranged in two pairs, side by side.
→ *Gynoecium* (inner whorl): made up of one or more carpels. Each carpel is a closed structure inside which one or more ovules develop. Each ovule develops from the nucellus and becomes surrounded by two integuments. The lower part of the carpel, which surrounds the ovules, is called the ovary and bears at its apex a stalk-like structure, the style. This ends in a receptive surface, the stigma. The diagram below shows the structure of a typical insect-pollinated flower.

Action point

Annotate the diagram of a typical flower describing the functions of the various parts.

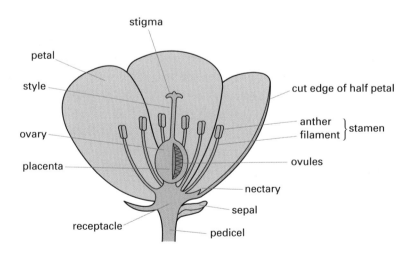

Checkpoint 4

What is the function of the tapetum layer in the pollen sac?

Pollen grain development

In the early stages of development, each pollen sac contains a mass of microspore mother cells, surrounded by a nutritive layer called the tapetum (see diagram opposite).

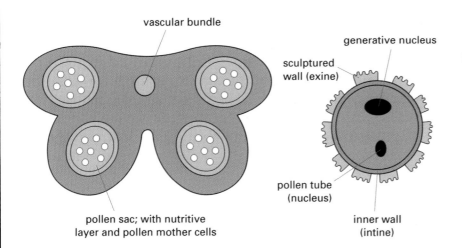

vascular bundle

generative nucleus

sculptured
wall (exine)

pollen tube
(nucleus)

pollen sac; with nutritive
layer and pollen mother cells

inner wall
(intine)

Action point

Draw a pollen sac in TS and LS as seen under the microscope.

Each microspore mother cell undergoes **meiosis** to give a tetrad (four) of haploid cells. Each of these cells develops into a pollen grain. A thick outer wall, the **exine**, and a thin inner wall, the **intine**, are secreted. Inside the pollen grain the haploid nucleus undergoes mitosis to produce a generative nucleus and a pollen tube nucleus. The generative nucleus will later give rise to the two male nuclei. As the pollen grains mature, changes take place in the anthers. The cells of the tapetum shrink and degenerate, and fibrous layers develop beneath the epidermis. When the pollen is ripe, the outer layers of the anthers dry out and tensions are set up in lateral grooves. Eventually dehiscence occurs and the edges of the pollen sacs curl away, exposing the pollen grains.

Links

The process of meiosis is dealt with on pages 162 and 163. Mitosis is dealt with on pages 28 and 29.

Checkpoint 5

How many nuclei develop in a pollen grain?

Ovule development

In the early stages of the development of an ovule, the nucellus arises as a tiny lump of tissue growing out from the placenta of the carpel. At its apex a megaspore mother cell develops and undergoes meiosis to form four haploid megaspores. Normally only one of these cells continues to develop, the rest abort. The single megaspore gets bigger, gaining its nutrition from the nucellus tissue. It undergoes three successive mitotic divisions and forms an embryo sac containing eight haploid nuclei. In the embryo sac the eight nuclei are arranged in a definite pattern. Three cells are situated at the micropyle end of the ovule, making up what is referred to as the egg apparatus, the central cell being the female gamete or egg cell. At the opposite end is another group of three cells called the antipodal cells. The remaining two nuclei, called the polar nuclei, migrate to the centre of the embryo sac.

Action point

Draw a labelled diagram showing a mature ovule.

Checkpoint 6

How many nuclei are produced by meiosis in a single megaspore?

Exam practice   answers: page 184

1. Describe the male and female parts of the flowering plant, giving a full account of the process of gametogenesis which takes place in them.

Pollination and fertilization

A key feature of the success of flowering plants is their relationship with animals. Pollen grains have no power of independent movement and have to be transferred to the female part of the flower to ensure fertilization. Flowering plants have evolved the strategy of attracting animals, particularly insects, to their flowers, feeding them and exploiting their mobility to transfer pollen from flower to flower. Some plants are pollinated by the action of wind.

Pollination

Pollination is necessary in order that the pollen grains, containing the male gametes, are brought into contact with the gynoecium so that fertilization can be achieved. This means that pollen grains must be transferred from the ripe anther to the receptive stigma.

Self-pollination

In some species self-pollination occurs and the pollen from the anthers of a flower need only be transferred to the stigma of the same flower or another flower on the same plant.

Cross-pollination

In a large number of species, cross-pollination is more normal, where pollen is transferred from the anthers of one flower to the stigma of another flower on another plant of the same species. There are a variety of mechanisms that ensure cross-pollination or prevent self-pollination, so ensuring outbreeding. Anthers and stigma may mature at different times, they may be at different levels in the flower or there may be separate male and female flowers on different plants.

There are two main methods of pollination:

→ by wind
→ by animal vectors such as insects, birds or mammals

Wind pollination

Wind-pollinated flowers tend to be inconspicuous, with small petals and sepals, lacking scent or nectaries. Either the stamens are very numerous or there are large anthers present, producing masses of light smooth pollen. The filaments are very long and dangle outside the flower so that the pollen is easily dispersed by air currents. The stigmas are long and feathery, providing a large surface area to trap the pollen. They also often hang outside the flower thereby catching pollen in air currents.

Insect pollination

Insect-pollinated flowers are large and brightly coloured to attract insects. They have coloured, often shiny, petals and may produce a scent and nectar. The scent attracts the insects and the nectar is collected for food. The stamens are few in number and are fixed inside the flower, where insects must brush past them in order to reach the nectar. The pollen

Examiner's secrets

The two forms of pollination have different genetic consequences. When answering a question on plant reproduction, be prepared for part of the question to ask you to compare the evolutionary significance of inbreeding and outbreeding.

Checkpoint 1

What is the difference between the definitions of self- and cross-pollination?

Checkpoint 2

List three methods of preventing inbreeding in the flowering plant.

Checkpoint 3

Why do wind-pollinated flowers produce enormous amounts of pollen?

Checkpoint 4

Why do wind-pollinated flowers have feathery stigmas which protrude outside the flower?

grains are large, often with elaborately sculptured walls, which help them stick to the bodies of the insects.

Successful pollination

Successful pollination results in compatible pollen grains reaching the stigma. If the pollen grains are incompatible, they will not germinate successfully. The stigma produces a sugary solution in which the pollen grains germinate, producing a fine pollen tube. This grows down into the tissues of the style, secreting enzymes which digest some of the cells in order to gain nutrients. The pollen tube nucleus is situated near the top of the tube, with the two male nuclei close behind.

Fertilization

The pollen tube grows through the micropyle and passes into the ovule and then to the embryo sac. As the pollen tube penetrates the embryo sac, the tip opens and the two male gamete nuclei enter, the pollen tube nucleus having disintegrated earlier.

→ The first male nucleus fuses with the female, or egg, nucleus to form a zygote.
→ The second male nucleus fuses with the two polar nuclei to form a triploid endosperm nucleus.

These events constitute double fertilization, a unique feature of the flowering plants (see diagram below).

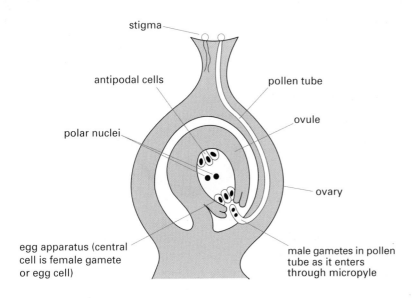

Test yourself

Construct a table comparing the features of a wind- and insect-pollinated flower. Include reasons for the various adaptations.

Action point

Construct a table of differences between wind- and insect-pollinated flowers.

Checkpoint 5

What is the difference between pollination and fertilization?

Watch out!

Be careful not to confuse which nuclei fuse together. Draw a diagram of fertilization and using matching colours, colour in the nuclei which fuse together.

Checkpoint 6

How is endosperm formed?

Checkpoint 7

How many megaspore nuclei are involved in fertilization?

Checkpoint 8

What is the importance of the triploid endosperm nucleus?

Exam practice   answers: page 184

1. Describe the way in which fertilization takes place from the time the pollen lands on the stigma.

Seed development and germination

The seed develops from the fertilized ovule and contains an embryonic plant and a food store. After a period of dormancy and when environmental factors are favourable stored food will be mobilized and the seed will germinate.

Development of the seed

- → The diploid zygote nucleus begins to divide by mitosis and eventually an embryo, consisting of a plumule, a radicle and one or two cotyledons, is formed.
- → The endosperm nucleus also divides mitotically, forming a large number of nuclei and eventually an endosperm tissue develops. These cells become filled with stored food materials, providing reserves for the developing embryo.
- → The integuments fuse together and become the seed coat or testa.
- → The carpel wall, the pericarp, becomes the fruit.
- → The ovule becomes the seed.

Structure of the seed

The diagram below shows the structure of a broad bean seed.

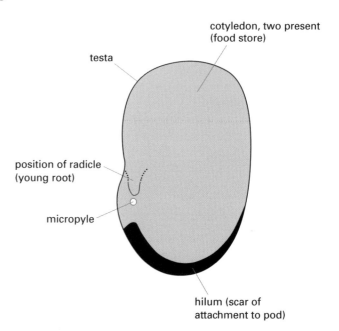

cotyledon, two present (food store)

testa

position of radicle (young root)

micropyle

hilum (scar of attachment to pod)

Germination

Germination of the seed usually occurs during a period of dormancy. The three main requirements for successful germination are:

- → a suitable *temperature* – this is related to enzyme activity
- → *water* – needed for the mobilization of enzymes, vacuolation of cells and for transport
- → *oxygen* – energy is needed for growth, production of ATP

Checkpoint 1

After fertilization what happens to the sepals and stamens of the flower?

Checkpoint 2

When the ovary forms, what will
(a) the ovary wall become?
(b) the ovule become?

Examiner's secrets

Check your specification for a named seed.

Checkpoint 3

Why is oxygen required for seed germination?

Mobilization of food reserves during germination

Food reserves in seeds are insoluble in water and cannot be transported in the seedling. They must be broken down into relatively simple soluble substances that dissolve in water so that they can be moved to the growing apices of the plumule and radicle. Water is taken up rapidly by the seed in the initial stages, causing the tissues to swell as well as mobilizing the enzymes. The testa, or seed coat, will rupture as the radicle pushes its way through. The radicle will grow downwards and the plumule upwards.

Secretion of hydrolytic enzymes, e.g. amylase, which hydrolyses starch into maltose, is triggered by the hormone gibberelic acid (GA) produced by the embryo. The GA diffuses to the aleurone layer surrounding the endosperm where protein synthesis occurs (see the diagram of a barley seed below).

Checkpoint 4

Why do seeds need a food store?

Checkpoint 5

Why does the seedling establish a root system as the first step in growth?

Checkpoint 6

Where is gibberelin produced?

Checkpoint 7

Name the layer of the seed where GA triggers protein synthesis.

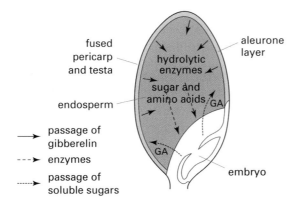

Links

Plant growth regulators are studied on pages 107 and 108.

The success of the angiosperms

The angiosperms are the dominant plant group on Earth. The rapid rate at which they achieved diversity may be attributed to several features.

→ Their sexual reproduction is rapid, the interval between flower production and setting of seed being usually a matter of weeks.
→ The protection provided by a closed ovary for the growth of the pollen tube also makes possible an efficient incompatibility system to exclude self-pollination.
→ The fertilization of gametes is independent of water.
→ The unique double fertilization ensures that a food store is only produced if the ovum is fertilized.
→ The cooperative venture between insects and flowering plants has been of mutual advantage.

Exam practice   answers: page 184

1. Give an account of the formation of seeds after fertilization and describe the germination of seeds.

Human reproduction

In the human reproductive system the gametes are produced in special paired glands called gonads. The male gametes or sperm are produced in the testes and the female gametes or egg cells in the ovaries. You need to study the development of gametes, which is known as gametogenesis.

Male reproductive system

The male system consists of a pair of testes, which produce spermatozoa; the penis, which is an intromittent organ; genital ducts connecting the two; and various accessory glands, which provide constituents for the semen. Each testis consists of about 1 000 seminiferous tubules, which produce the spermatozoa. The seminiferous tubules also contain interstitial cells, which produce the male hormone testosterone. When sperm have been produced they collect in the vasa efferentia and then pass to the head of the epididymis, where they mature. They then pass along the coiled tube to the base of the epididymis, where they are stored for a short time before passing via the vas deferens to the urethra during ejaculation.

Action point

Draw a diagram showing the structure of a testis.

Female reproductive system

There are two ovaries, each of which contains ova, which are produced in the germinal epithelium. Once formed, the ova develop in follicles. Mature follicles migrate back to the surface when their development is complete, so that the ova can be shed. The structure of the ovary is shown in the diagram below.

Action point

Annotate the diagram of the ovary.

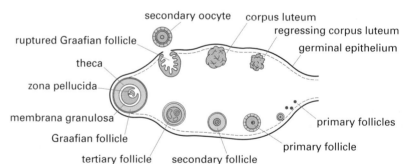

secondary oocyte   corpus luteum
ruptured Graafian follicle   regressing corpus luteum
    germinal epithelium
theca
zona pellucida
membrana granulosa   primary follicles
Graafian follicle   primary follicle
tertiary follicle   secondary follicle

Ova are passed to the fallopian tube or oviduct, which conveys them to the uterus or womb. The uterus is a muscular organ that opens into the vagina.

Gametogenesis

The cells of the germinal epithelium of both the testis and the ovary undergo a sequence of mitotic and meiotic divisions to form gametes. First there is a multiplication stage which involves repeated mitotic divisions to produce spermatogonia and oogonia. Once formed they grow to full size and then undergo maturation, which involves meiotic division and then differentiation into the mature gametes. The diagram opposite shows the sequence of events in **gametogenesis**.

Watch out!

Make sure you know at which stage meiosis I and meiosis II take place.

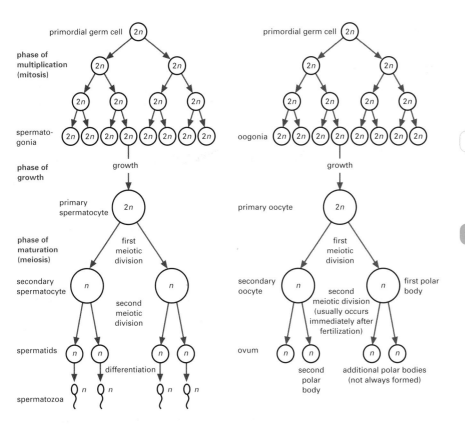

Links

Details of mitosis are on pages 28 and 29. Details of meiosis are on pages 162 and 163.

Checkpoint 1

Construct a table to show the differences between the steps in gametogenesis in the male and female.

Formation of spermatozoa

These are produced in the germinal epithelium of the seminiferous tubule and are nourished and protected by the Sertoli cells. They form primary spermatocytes, which then form haploid secondary spermatocytes after the first meiotic division. After the second meiotic division they form spermatids which differentiate into spermatozoa.

Formation of ova

Oogonia, which are formed before birth, undergo mitosis to form primary oocytes, which then mature just before ovulation. About 2 million of these are formed in the fetal ovary but only about 450 will later develop into secondary oocytes. Before ovulation, a primary oocyte undergoes the first meiotic division to form the haploid secondary oocyte and its first polar body. The secondary oocyte begins the second meiotic division, but this is arrested at metaphase unless fertilization takes place. On fertilization this division is completed to form a large ovum and a second polar body. Once this division has taken place the nucleus of the ovum fuses with that of the sperm to form a zygote, which will then develop into an embryo.

Checkpoint 2

What type of cell division takes place in the formation of spermatogonia?

Checkpoint 3

How many spermatozoa are produced from eight spermatogonia?

Checkpoint 4

How many ova are produced from eight oogonia?

Exam practice   answers: page 185

1. Give an illustrated account of the anatomy of the male reproductive system of a mammal indicating the functions of the structures you have described.

Hormonal control of reproduction

In all mammals female sexual activity is cyclical. A cycle in the ovary leads to the production of mature ova and a uterine cycle prepares the uterus to receive fertilized eggs. In humans there is a regular 28-day cycle controlled by hormones secreted by the pituitary gland and the ovary.

Menstrual cycle

The events that occur in the course of the menstrual cycle follow a set pattern that is regulated by hormones from the pituitary gland and ovaries. If fertilization occurs, the normal cyclical pattern is interrupted and an additional source of hormones, the placenta, comes into play. The main hormones involved, and their functions, are as follows.

→ *Follicle stimulating hormone* (FSH) secreted by the anterior pituitary:
  → causes Graafian follicles to develop in the ovary
  → stimulates the tissues of the ovary to produce oestrogen
→ *Oestrogen* produced by the ovary:
  → inhibits the production and release of FSH
  → causes repair of the uterine wall following menstruation
  → builds up in concentration during the first two weeks of the menstrual cycle until it stimulates the pituitary gland to produce luteinizing hormone
→ *Luteinizing hormone* (LH) secreted by the anterior pituitary:
  → brings about ovulation
  → causes a Graafian follicle to develop into a corpus luteum which produces oestrogen and progesterone
→ *Progesterone and oestrogen:*
  → inhibit FSH and LH production and therefore stop further follicle development
  → progesterone only causes development of the uterine wall prior to implantation

If fertilization does not take place, the falling levels of FSH and LH allow the corpus luteum to degenerate and the concentrations of progesterone and oestrogen also fall. As a result FSH and LH production is no longer inhibited. The thickened epithelium of the endometrium is shed, producing the menstrual flow.

If fertilization occurs, progesterone and oestrogen, produced initially by the corpus luteum and later by the placenta, are present in high concentration. LH production and then ovulation are inhibited; follicle development and production of oestrogen cease and the uterus lining is maintained.

Action point

Draw graphs showing changes in the levels of hormones involved in the menstrual cycle.

Action point

Rephrase the list opposite into full paragraphs describing the hormonal control of the menstrual cycle.

Checkpoint 1

Explain how progesterone, with smaller amounts of oestrogen, maintains the receptive state of the uterine wall and prevents ovulation.

Checkpoint 2

On what day of the cycle does ovulation occur?

Hormone regulation in the male

There is no sexual cycle in the male but gonadotrophic hormones are produced from the anterior lobe of the pituitary gland:

→ *FSH* stimulates sperm development
→ *LH* stimulates the production of testosterone
→ *testosterone* controls the development of male reproductive organs and secondary sexual characteristics

Checkpoint 3

Which organ produces testosterone?

Exam practice   answers: page 185

1   The diagram below shows the relationship between:

   (i)   the pituitary gonadotrophins
   (ii) the ovarian steroids
   (iii) follicle development
   (iv) the thickness of the endometrium during the human oestrous cycle

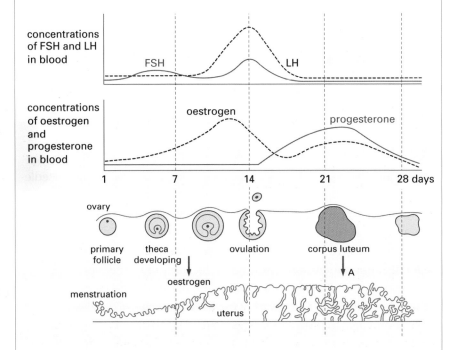

Examiner's secrets

Questions on hormonal control often ask for the analysis of graphs. Many of the answers can be deduced from information provided. Study the graph carefully; always look for peaks and troughs, which mark significant events. Read the whole question over at least once thoroughly before starting to write your answer.

   (a) From which cells is the hormone oestrogen secreted and what evidence is there for this in the diagram?
   (b) Give one effect of the hormone oestrogen on the pituitary gland. Explain your answer with reference to the diagram.
   (c) What is represented by the arrow A on the diagram?
   (d) What evidence is there from the diagram that the hormone progesterone is involved in a negative feedback mechanism?
   (e) How is a knowledge of the effects shown in the diagram involved in the formulation of the contraceptive pill?

2   (a) Name one hormone involved in birth and state its effect.
   (b) What is the role of hormones in lactation?

Fertilization and development

Internal fertilization ensures that the sperm are deposited in the female's reproductive tract. When a sperm fuses with an egg cell a zygote is formed and this develops by mitosis to form a mass of cells. The zygote is implanted in the uterus wall where it continues to develop. You need to study the placenta and describe its role in the development of the embryo until birth.

Fertilization

Sperm are deposited near to the cervix and quickly pass to the top of the fallopian tube by contractions of the uterus and fallopian tube. Sperm are highly fertile for 12–24 hours after release into the female tract but can fertilize an ovum only after a process called capacitation, which takes several hours. It involves changes in the membrane covering the **acrosome**, a thin cap over the nucleus. The structure of the sperm is shown in the diagram below.

Checkpoint 1

Why does the sperm have large numbers of mitochondria in the middle piece?

acrosome   axial filament in tail

nucleus   axial filament in neck   mitochondrion   tail sheath

When the sperm reach an oocyte, contact with the jelly coat results in the acrosome membrane rupturing and enzymes are released. The enzymes soften the layers of cells (corona radiata and zona pellucida) surrounding the oocyte. Inversion of the acrosome results in a fine needle-like filament developing at the tip of the sperm and this pierces the already softened portion of the membrane. The whole process is called the **acrosome reaction** and it enables the sperm to penetrate the egg. This entry stimulates reactions of the oocyte that brings about the formation of the **fertilization membrane**, preventing the entry of further sperm. Entry of the sperm also stimulates the completion of the second meiotic division of the oocyte nucleus. The nuclei of the ovum and sperm are drawn together and fuse to form a diploid nucleus. The fertilized ovum is known as the zygote and it begins to divide by mitosis to form the embryo.

Checkpoint 2

What is the importance of the fertilization membrane?

Implantation

As soon as the ovum is fertilized it begins to divide until a hollow ball of cells, the blastocyst, is produced. The development of the zygote continues during its passage down the fallopian tube, with the outer layer of cells of the blastocyst developing into the embryonic membranes, the chorion and amnion. After reaching the uterus the blastocyst embeds in the endometrium. This is called implantation.

Examiner's secrets

You do not need to learn about embryonic development in any detail apart from the role of the amnion.

The placenta

The placenta develops in the region where implantation occurred. The chorion develops villi, which grow into the surrounding uterine tissue from which they absorb nutrients. These villi form part of the placenta, which is connected to the fetus by the umbilical cord. The amnion develops a membrane around the fetus and encloses the amniotic fluid, which protects the fetus by cushioning it from physical damage.

Functions of the placenta

The placenta has several important functions.

→ To nourish the developing fetus: in the placenta the maternal blood is brought close to that of the fetus but they do not mix. Finger-like projections of the allanto-chorion grow into the endometrium and become bathed by maternal blood. There is a minimal barrier over a huge surface area, allowing the diffusion and active transport of materials across the placenta. Water, amino acids, glucose, lipids, mineral salts, simple proteins, vitamins, hormones, antibodies and oxygen pass from the mother to the fetus.

→ The removal of wastes: carbon dioxide, urea and other nitrogenous wastes are transferred to the maternal blood to allow their excretion by the mother.

→ To act as a depressurizing unit: it protects the fetal circulation from the high maternal blood pressure.

→ It acts as a filter: the placenta prevents the passage of maternal hormones that could affect fetal development, while allowing some hormones across. It also prevents the passage of some pathogens, although some viruses and bacteria can cross.

→ During pregnancy the placenta progressively takes over the role of hormone production.

Checkpoint 3

Why is it important that the blood circulation of the mother and fetus do not mix?

Examiner's secrets

You may be required to study the fetal circulation. Check your specification.

Links

The menstrual cycle is discussed on page 158.

Birth

During pregnancy the quantity of different hormones produced changes, with the amount of oestrogen increasing and that of progesterone decreasing. This triggers the birth process and the hormone oxytocin is produced from the posterior lobe of the pituitary gland. This hormone causes contractions of the uterus, which results in the birth of the baby, followed a little later by the placenta, or afterbirth.

Lactation

After the birth the anterior lobe of the pituitary gland produces the hormone prolactin, which causes the mammary glands to produce milk.

Checkpoint 4

What inhibits prolactin production?

Exam practice answers: page 185

1. Give an account of the functions of the placenta.

Meiosis

Meiosis occurs in the formation of gametes. It is referred to as a reduction division because the pairs of chromosomes in diploid organisms are separated in the formation of haploid gametes.

Links

Mitosis is studied on pages 28 and 29.

Checkpoint 1

Distinguish between the haploid and diploid condition.

Checkpoint 2

Explain why chromosomes occur in homologous pairs in diploid cells.

The jargon

In *homologous* pairs of chromosomes, one of the pairs is inherited from the father and the other from the mother.

Checkpoint 3

What is a chiasma?

The process of meiosis

Meiosis is the reduction division that occurs during gamete formation in sexually reproducing organisms. In this division the diploid number of chromosomes ($2n$) is reduced to the haploid (n).

Usually there are two cycles of division:

→ firstly, where the chromosome number is reduced
→ secondly, where the two new haploid nuclei divide again in a division identical to that of mitosis

The net result is that four haploid nuclei are formed from the parent nucleus. Like mitosis, meiosis is a gradual process but for convenience it is divided into the four phases, prophase, metaphase, anaphase and telophase, these phases occurring once in each of the two divisions.

Prophase 1

This stage is similar to prophase in mitosis in that the chromosomes shorten and fatten and become visible but in meiosis they associate in their homologous pairs, and each pair is called a bivalent. At a certain stage in prophase 1 the chromosomes appear double-stranded, as the sister **chromatids** of each chromosome separate and become visible. These chromatids wrap around each other and then partially repel each other but remain joined at certain points called chiasmata. At these points chromatids may break and recombine with a different but equivalent chromatid. This swapping of pieces of chromosomes is called crossing over and is a source of genetic variation (see diagram below).

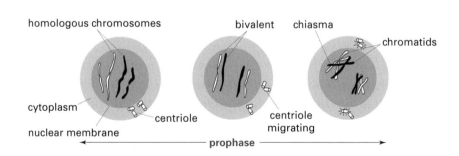

Metaphase 1

At this stage when the pairs of homologous chromosomes arrange themselves on the equator of the spindle they do so randomly. This random distribution and consequent independent assortment of chromosomes produces new genetic combinations. At this stage, when the pairs of homologous chromosomes align themselves on the equator of the spindle, the maternal and paternal chromosomes are arranged randomly.

Anaphase 1

Homologous chromosomes are pulled apart: one of each pair is pulled to one pole, its sister chromosome to the opposite pole. Thus each pole receives only one of each homologous pair of chromosomes and because of their random arrangement at metaphase these will be a mixture of maternal and paternal chromosomes. This is called independent assortment of chromosomes and produces new genetic combinations. The chromosomes reach the opposite poles and the nuclear envelope reforms (see diagram below).

Telophase 1

The nucleus may enter interphase but in some cells this stage does not occur and the cell passes from anaphase 1 directly into prophase 2. Meiosis II is a typical mitotic division and the result is the formation of four haploid daughter cells.

Action point

Construct a table of differences between mitosis and meiosis.

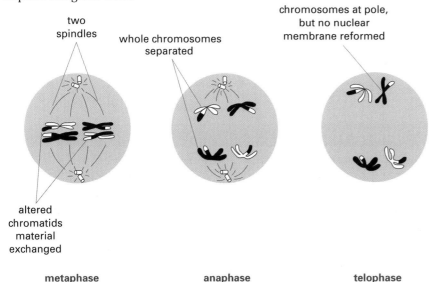

two spindles

whole chromosomes separated

chromosomes at pole, but no nuclear membrane reformed

altered chromatids material exchanged

metaphase   anaphase   telophase

Checkpoint 4

Why do gametes formed during meiosis have different chromosome contents?

Examiner's secrets

The sources of variation are central to the concept of evolution.

The significance of meiosis

In the long term, if a species is to survive in a constantly changing environment and to colonise new environments, sources of variation are essential. There are three ways of creating variety:

→ Each of the chromosomes making up a homologous pair carries different genetic material. During sexual reproduction the **genotype** of one parent is mixed with that of the other when haploid gametes fuse.

→ The different pairs of **homologous chromosomes** arrange themselves on the spindle during metaphase 1 of meiosis. When they subsequently separate they do so entirely independently of each other, so that the daughter cells contain different combinations of chromosomes.

→ Crossing over during chiasmata formation during prophase 1 of meiosis. Equivalent parts of homologous chromosomes may be exchanged thus producing new combinations and the separation of linked genes.

The jargon

Genotype means the genetic make-up.

Exam practice   answers: page 186

1 Write a detailed account of the process of meiosis in plants (the biological significance of the stages within the process is not required).

Variation and evolution

Sexual reproduction produces genetic variation amongst individuals in a population. In the long term, if a species is to survive in a constantly changing environment and to colonise new environments sources of variation are essential. The genotype of an organism gives it the potential to show a particular characteristic. The degree to which the characteristic is shown is also affected by the organism's environment.

Types of variation

→ Most characters are controlled by a number of genes and the differences in the character are not clear-cut. A character within a population which shows a gradation from one extreme to another shows **continuous variation**, for example, height.
→ Characters that are clear-cut and easy to tell apart are controlled by a single gene. When these types of difference occur it is known as **discontinuous variation**. This gene may have two or more alleles. There are no intermediate types, for example, light and dark forms in some moth species.

Origins of variation

Non-heritable variation

The environment has a role in determining phenotypic variation. Environmental factors in humans may include diet and exercise whereas plants are affected by temperature, light and available nutrients. Variations due to the effect of the environment have little evolutionary significance as they are not passed from one generation to the next.

Heritable variation

An organism will inherit genes, giving it a theoretical maximum size, but whether or not this is reached will depend upon nutrition during the growth period and other environmental factors. Thus, if organisms of identical genotype are subject to different environmental influences, they show considerable variety. Because these influences are varied, they are largely responsible for continuous variation in a population.

Checkpoint 1

Distinguish between continuous and discontinuous variation.

Checkpoint 2

Is the presence or absence of ear lobes an example of continuous or discontinuous variation?

Overproduction

All organisms have the reproductive potential to increase their populations, although they rarely do so. As a population increases, various environmental factors come into play to keep the numbers down. Organisms must compete for limited resources. For example, plants compete for light, space, mineral ions, etc.; animals compete for food, shelter, etc. The number of young produced is far greater than the number which will survive to become adults. Many young die before maturity and so do not reproduce. Two types of competition are apparent:

→ Competition between individuals of the same species (intra-specific competition). This is the basis of the origin of species by natural selection.

→ Competition between individuals of different species (inter-specific competition) is illustrated by predator–prey relationships.

Selection pressure

What determines which individuals die and which survive? Is it a matter of luck or are some individuals born with a better chance of survival than others? Variation within a population of organisms means that some will have characteristics which give them an advantage in the 'struggle for survival'.

In rabbits coat colour may vary. Most rabbits have alleles which give the normal brown colour. A small number may be homozygous for the recessive allele which gives a white coat. A white rabbit will stand out and is more likely to be killed by a predator. As the white rabbit is unlikely to survive to become a mature adult, the chances of it reproducing and passing on its allele for white coat are very small. The allele for white coat will remain rare in the population. Predation by foxes is an example of a selection pressure and increases the chances of some alleles being passed on to the next generation, and decrease the chance of others. The effects of such selection pressures on the frequency of alleles in a population is called natural selection. These better adapted organisms are more likely to pass on their characteristics to succeeding generations. The organism's environment exerts a **selection pressure** and this determines the spread of any allele within the gene pool.

Checkpoint 3

Apart from mutations, list three sources of variation.

Checkpoint 4

What is meant by the term selection pressure?

The jargon

Selection, in the context of evolution, is the process by which organisms that are better adapted to their environment survive and breed, while those less well adapted fail to do so.

Checkpoint 5

What is a somatic mutation?

Links

See also gene and chromosome mutations on pages 170 and 171.

Exam practice   answers: page 186

1 Using the information in the table below, answer the questions:

Trait	Average difference between pairs of:		
	Identical twins reared in the **same** family	Non-identical twins reared in the **same** family	Identical twins reared in **different** families
Height	18 mm	44 mm	18 mm
Weight	1.86 kg	4.5 kg	4.4 kg
Head width	2.85 mm	4.2 mm	2.85 mm
IQ		5.9	9.9 8.2

(a) What is the effect of moving an identical twin to a different family? Explain your answer.

(b) Suggest why data were collected for non-identical twins reared together. Explain fully what these data show.

Mendel's laws

Mendel carried out the first documented scientific investigation of inheritance and formulated two laws, which are basic to the science of genetics. You need to understand the principles of monohybrid and dihybrid Mendelian inheritance.

Check the net

You'll find up-to-date information on Mendelian inheritance at www3.ncbi.nlm.nih.gov/Omim/searchomim.html

Examiner's secrets

You should learn the definitions of all the genetic terms.

Checkpoint 1

Distinguish between a gene and an allele.

Watch out!

Don't confuse genotype and phenotype.

Checkpoint 2

What is meant by self-pollination?

Action point

Describe the process of a dihybrid cross from the planting of the parent seeds to the gathering of data in the F2 generation.

Don't forget

A 'back-cross' is a method used in genetics to determine whether a particular dominant characteristic observed in an organism is determined by one or two dominant alleles.

First law

Mendel's early experiments were based on selecting pea plants of two varieties that showed clearly separable characteristics, such as tall and dwarf plants and round and wrinkled seeds. From his results he formulated the law of segregation, which states that 'the characteristics of an organism are determined by factors which occur in pairs. Only one member of a pair of factors can be represented in any gamete'. The inheritance of a single pair of contrasted characters is known as monohybrid inheritance.

Genetic terms

→ *Dominant* allele: of the pair of alleles on homologous chromosomes, the one that always produces an effect on the appearance of the organism when present is the dominant allele. It is usually represented by a capital letter, e.g. T (for tall).
→ *Recessive* allele: the allele that produces an effect only when present as an identical pair is the recessive allele, e.g. t.
→ *Genotype*: the combination of alleles found in an individual.
→ *Phenotype*: the appearance of an organism, determined by the genotype.
→ *Homozygous*: if both alleles are the same, e.g. TT, tt.
→ *Heterozygous*: if the pair of alleles are dissimilar, e.g. Tt.
→ F1: first filial generation; F2, second filial generation.

In monohybrid crosses, two heterozygous individuals will produce a phenotypic ratio of three dominant to one recessive.

Second law

Mendel also investigated what happened if he used plants which differed by having two pairs of contrasting characters. In one of his experiments he investigated the inheritance of seed shape (round vs. wrinkled) and seed colour (green vs. yellow). He crossed homozygous plants with round and yellow seeds with homozygous plants with wrinkled and green seeds. When plants grown from these seeds were self-pollinated the seeds produced were of four different types of shape and colour of seed coat. The four types were:

→ round yellow   → round green
→ wrinkled yellow   → wrinkled green

He also found that these types were in the ratio of approximately 9:3:3:1, proportions now known as the dihybrid ratio. This led him to formulate his second law, stating that 'either one of a pair of characteristics may combine with either of another pair'.

The probability of the four alleles appearing in any of the F2 offspring is:

→ round (dominant) $^3/_4$ → yellow (dominant) $^3/_4$
→ wrinkled (recessive) $^1/_4$ → green (recessive) $^1/_4$

The probability of combinations of alleles appearing in the F2 is:

→ round and yellow = $^3/_4 \times {}^3/_4 = {}^9/_{16}$
→ round and green = $^3/_4 \times {}^1/_4 = {}^3/_{16}$
→ wrinkled and yellow = $^1/_4 \times {}^3/_4 = {}^3/_{16}$
→ wrinkled and green = $^1/_4 \times {}^1/_4 = {}^1/_{16}$

Below is a genetic diagram to explain the results which Mendel obtained, represented in a Punnett square.

R represents round seed (dominant)   r represents wrinkled seed (recessive)
Y represents yellow seed (dominant)   y represents green seed (recessive)

parental phenotypes	pure breeding round yellow			pure breeding wrinkled green
parental genotypes (2n)	RRYY			rryy
gametes (n)	all RY			all ry
F1 genotype (2n) random fertilization self-fertilization		all RrYy RrYy × RrYy		
gametes identical from ovules and pollen grains	RY	Ry	rY	Yy

♀\♂	RY	Ry	rY	ry
RY	RY RY round yellow	Ry RY round yellow	rY RY round yellow	ry RY round yellow
Ry	RY Ry round yellow	Ry Ry round green	rY Ry round yellow	ry Ry round green
rY	RY rY round yellow	Ry rY round yellow	rY rY wrinkled yellow	ry rY wrinkled yellow
ry	RY ry round yellow	Ry ry round green	rY ry wrinkled yellow	ry ry wrinkled green

F2 genotypes (2n) and phenotypes shown in a Punnett square

9 round yellow
3 round green

3 wrinkled yellow
1 wrinkled green

Exam practice   answers: page 186

1 Using two named independent contrasting pairs of alleles, show
   diagrammatically a cross between a homozygous dominant individual
   and a homozygous recessive individual. With diagrams, show the results
   of crossing an individual from the F1 generation with:
   (a) an individual with an identical genotype
   (b) an individual with only recessive genes
   Indicate the proportions of the phenotypes in the F2 in each case.

Watch out!

Be sure that you understand what is meant by 'probability'. If Mendel's experiment produced 320 seeds in the F2 generation the probability of there being wrinkled and green seeds would be 320 × 1/16 = 20.

Checkpoint 3

Why was Mendel fortunate in the choice of the pea plant for his experiments?

Action point

Use the Chi^2 test to test the significance of the difference between observed and expected results.

Watch out!

Use the Punnett square to minimize the risk of making mistakes.

Action point

Colour in the Punnett square using a different colour for each of the four phenotypes.

Checkpoint 4

If Mendel crossed homozygous plants having round and yellow cotyledons with homozygous plants having wrinkled and green cotyledons and he obtained 608 seeds in the F2 generation, how many seeds would have wrinkled and green cotyledons?

Examiner's secrets

Once you have practised a number of crosses it is very easy to start missing out stages or explanations and to concentrate on achieving the outcome only. This may make what you have written down in an exam difficult to follow. Even if you achieve the expected outcome, you may not gain full credit. Always carry out genetic crosses in full, giving concise explanations.

Deviations from Mendel's laws and sex linkage

Not all the characteristics are controlled by single genes which behave independently, as was the case in Mendel's experiments. You need to know about incomplete or co-dominance, and also how sex is determined and how genes on sex chromosomes cause unexpected ratios.

Examiner's secrets

You are only required to study co-dominance in monohybrid crosses.

Watch out!

Co-dominance may be expressed using the same letter throughout, e.g. Rr, or by using different letters to represent each allele, e.g. R (red), W (white), RW (pink) in the case of antirrhinum.

Checkpoint 1

If roan bulls and cows are mated together what genotypes and phenotypes would you expect in the offspring?

Co-dominance

When alleles express themselves equally in the phenotype this is known as co-dominance. In most cases the heterozygote shows a phenotype intermediate between those of the two homozygotes. Examples of co-dominance are:

→ antirrhinums – red (RR), pink (RW) and white (WW) flowers
→ shorthorn cattle – red (RR), roan (RW) and white (WW) coat colour

The genetic diagram for these crosses is the same as that illustrating Mendel's first law but in the F1 all individuals have the intermediate phenotype.

Multiple alleles

Many characteristics in organisms are controlled not by pairs of alleles but by three or more alleles of which any two may occupy the gene loci on the homologous chromosomes. An example of sets of phenotypes controlled by multiple alleles is the ABO blood groups in humans.

Epistasis

Epistasis is a form of gene interaction in which a dominant allele at one locus modifies or masks the expression of an allele at a second locus. This can give rise to unusual ratios in genetic crosses. Examples of epistasis are inheritance of certain coat colours in mice and banding in the shell of the snail *Cepaea nemoralis*.

Sex determination

Most sexually reproducing animals show two morphologically distinct types, male and female, which are associated with the chromosomes found in the two types. One pair, known as the sex chromosomes, is similar in one sex and dissimilar in the other. The non-sex chromosomes are known as autosomes. In humans the male has dissimilar chromosomes, called X and Y, while the female has two similar X chromosomes.

Checkpoint 2

Draw a genetic diagram using XX female and XY male as the parental genotypes. What is the expected ratio in the offspring?

Sex-linked inheritance

The Y chromosome is much smaller than the X and carries very few genes. Therefore in the male any recessive genes carried on the X chromosome will express themselves in the phenotype. This is because they are unpaired and so there is no dominant gene present. This special form of inheritance is known as sex-linkage, an important feature of which is that the male cannot hand on the gene to his sons as they must receive the Y chromosome to become male. On the other hand, all his daughters must receive the recessive gene from him. Females who are heterozygous for sex-linked recessive traits are known as carriers and have a 50% chance of handing on the recessive gene to their sons. To obtain an affected (homozygous recessive) female, the father must be affected and the mother either affected or a carrier. Examples of sex-linked traits in humans are red/green colour-blindness and haemophilia. The process of sex-linked inheritance is shown in the diagram below.

Examiner's secrets

You are not required to study autosomal linkage.

Watch out!

Don't make the mistake of thinking that sex-linked characters only affect the male.

Checkpoint 3

What are the genotypes of the parents who produce a colour-blind female?

parental phenotypes	normal female		× colour-blind male		carrier female		× normal male	
parental genotypes (2n)	XX	×		X^cY	X^cX	×		XY
gametes (n)	X		X X^c	Y	X^c	X X		Y
F1 genotypes (2n)	XX^c		XX^c XY	X^cY	X^cX	XX	X^cY	XY
F1 phenotype	carrier females		normal male	colour-blind male	carrier female	normal female	colour-blind male	normal male

parental phenotypes	carrier female		× colour-blind male		colour-blind female		× normal male	
parental genotypes (2n)	X^cX	×		X^cY	X^cX^c	×		XY
gametes (n)	X^c		X X^c	Y	X^c	X^c X		Y
F2 genotypes (2n)	X^cX^c		XX^c X^cY	XY	X^cX	X^cX X^cY		X^cY
F2 phenotype	colour-blind female	carrier female	colour-blind male	normal male	carrier females		colour-blind males	

Exam practice   answers: page 187

1 (a) State two reasons why female haemophiliacs are rarely found in a population.

   (b) Using suitable symbols complete the genetic diagram to show the way in which haemophilia is inherited in the family shown.

Parental phenotype	*Carrier*	*Haemophiliac*
	mother	*father*
Parental genotype	....	....
Gametes		
Possible offspring genotypes		
Offspring phenotypes		

Mutations

A mutation is an unpredictable change in the genetic material of an organism. Mutations create completely new alleles and are an important source of genetic variation. There are two types of mutation, gene mutations and chromosome mutations.

Mutations

A mutation is a change in the amount, arrangement or structure in the DNA of an organism.

→ It may affect a single gene or a whole chromosome.
→ Most mutations occur in somatic (body) cells.
→ Only those mutations which occur in the formation of gametes can be inherited.
→ Mutations are natural, spontaneous, random events which may provide a source of material for natural selection pressures and therefore evolution.
→ Mutations rates are normally very low, therefore mutation has less impact on evolution than other sources of variation.
→ In general, organisms with short life cycles and more frequent meiosis show a greater rate of mutation.
→ The rate of mutations occurring can be increased by ionising radiation and **mutagenic** chemicals.

Mutations can happen in two ways:

→ DNA is not copied properly before cell division.
  Sometimes mistakes are made in the copying process so that new chromosomes are faulty. Usually they are small errors, involving only one gene, so they are called **gene mutations** or point mutations. However, this can be a serious problem for the individual if a very important gene is affected.
→ Chromosomes are damaged and break.
  If chromosomes break, they will normally repair themselves (the DNA will rejoin), but they may not repair themselves correctly. This can lead to large changes in the structure of the DNA, and may affect a large number of genes. These are called **chromosome mutations**.

Sickle-cell anaemia

A gene mutation (substitution) in the gene producing haemoglobin results in a defect called **sickle-cell anaemia**. The replacement of just one base in the DNA molecule results in the wrong amino acid being incorporated into two of the polypeptide chains which make up the haemoglobin molecule. The abnormal haemoglobin causes red blood cells to become sickle-shaped, resulting in anaemia and possible death. Haemoglobin S is produced instead of normal haemoglobin by a single base chain that causes valine to be substituted for glutamic acid. The mutant gene is co-dominant. In the homozygous state the individual suffers the disease but in the heterozygous state the individual has 30–40 per cent sickle cells, the rest are normal. The heterozygous condition is referred to as sickle-cell trait.

Checkpoint 1

Suggest why bacteria are often used in mutation experiments.

Checkpoint 2

Explain why gene mutations rarely show up in the phenotype.

Checkpoint 3

What are the chances of two parents with sickle-cell trait having a child with sickle-cell anaemia?

Checkpoint 4

What is the connection between sickle-cell trait and malaria?

Don't forget

Mutations happen naturally but the mutation rate is increased if organisms are exposed to mutagens. It is incorrect to state that mutagens cause cancer. Increased exposure to mutagens increases the **rate** of mutations occurring.

Non-disjunction

Non-disjunction is a process in which faulty cell division means that one of the daughter cells receives two copies of a chromosome while the other gets none. In **Down's syndrome** chromosome number 21 is affected. If this happens in an ovary, it results in an oocyte with either no chromosome 21 or with two copies instead of one. Oocytes with no chromosome 21 die but those with two copies survive and may be fertilized. The resulting zygote has three chromosome 21s with a total of 47 chromosomes. This condition is known as trisomy 21 and the zygote will develop into a child with Down's syndrome.

Why are mutations important?

Mutations are important because they increase variation in a population. Most mutations are harmful to the organism concerned. Beneficial mutations are very rare, but they may give a selective advantage to an organism.

→ These mutations cause sudden and distinct differences between individuals. They are therefore the basis of discontinuous variation.
→ There are potential advantages from mutations that are beneficial and may increase variation. However, most mutations are recessive to the normal allele. A recessive mutant allele must await replication in the gene pool over many generations before chance brings recessive alleles together, resulting in their expression.

Test yourself

Draw diagrams to show structural changes in chromosome mutation.

Checkpoint 5

Why can't somatic mutations be transmitted to the next generation?

Carcinogens

Substances that cause cancer are called carcinogens. These affect the DNA in cells, resulting in mutations. Mutations that occur in body or somatic cells often have no effect on an organism. Most mutated cells are recognized as foreign by the body's immune system and are destroyed. Occasionally the mutation may affect the regulation of cell division. Cancers are thought to start when changes take place in these genes. The term for a mutated gene that causes cancer is an oncogene. If a cell with such a mutation escapes the attack of the immune system, it can produce a lump of cells called a tumour. Tumours are usually harmless or benign, but sometimes the tumour cells are able to spread around the body and invade other tissues. This type of tumour is described as malignant, and the diseases caused by such tumours are cancers.

→ If a mutation is in a body cell, it may cause cancer. For example, increased exposure to UV light is linked to skin cancer.
→ If the mutation is in a gamete, it will not affect the individual producing the gamete, but will affect the zygote that develops from it, that is, the offspring.

The jargon

A mutagen which causes cancer is a carcinogen. X-rays, gamma radiation and UV light chemicals, such as polycyclic hydrocarbons in cigarette smoke, are mutagens.

Exam practice   answers: page 187

1. Modern bread wheat is thought to have evolved from *Triticum durum* ($2n = 28$) which developed from sterile hybrid Z formed from a cross between *Triticum monococcum* ($2n = 14$) and *Aegilops speltoides* ($2n = 14$), a wild grass. Explain fully why hybrid Z is sterile whereas *Triticum durum* is fertile.

Human Genome Project

Human DNA consists of about 3000 million bases. The Human Genome Project began in 1990 involving scientists in many countries. It has been a mammoth task determining the order of bases in the human genome as well as the identification of genes, their sequencing and mapping. The identification of the 20,000 to 25,000 genes in human DNA took 13 years to complete. The next challenge is to assign functions to the identified genes.

Aims of the project

The aims of the project were to:

→ determine the sequence of the four bases, A, T, G and C, throughout all human DNA
→ identify all the genes formed by the bases
→ find the location of the genes on the 23 human chromosomes
→ store this information on databases
→ consider all the ethical, social and legal issues which arise from obtaining information about the human genome

Beneficial applications of the project

The information enables scientists to know exactly which sections of DNA, on which chromosomes, are responsible for the many different inherited diseases. In gene tests, scientists scan a patient's DNA sample for mutated sequences.

→ For some types of gene tests, researchers design short pieces of DNA called **probes**, whose sequences are complementary to the mutated sequences. These probes will seek their complement among the three billion base pairs of an individual's genome. If the mutated sequence is present, the probe will bind to it and flag the mutation.
→ Another type of DNA testing involves comparing the sequence of DNA bases in a patient's gene to a normal version of the gene.
→ The main uses of genetic testing are in carrier screening, pre-implantation genetic diagnosis, newborn baby screening, presymptomatic testing for predicting adult-onset disorders such as Huntington's disease, presymptomatic testing for estimating the risk of developing adult-onset cancers and Alzheimer's disease, confirmation that an individual has a suspected disease and in forensic/identity testing.

Using a sample of DNA it is possible:

→ to identify whether that person is a carrier of a faulty gene, e.g. that which causes cystic fibrosis. People who are carriers may decide not to have children or to have an antenatal genetic test to check if their child will be born with the disease.

The jargon

A *genome* comprises the DNA in an organism, including its genes. Genes carry the information for making all the protein required by an organism. These proteins determine, amongst other things, the appearance of an organism, how it fights infection, etc.

Grade booster

The Human Genome Project is an area of intense research and would be expected to appear on an exam paper.

The jargon

Carrier screening involves identifying unaffected individuals who carry one copy of a gene (recessive) for a disease that requires two copies for the disease to be expressed.

Grade booster

Expansion of the pros and cons of gene testing should be researched. What are your views on genetic screening? Identify genes that play a contributory role in diseases.

→ to identify genes that play a contributory role in diseases, such as Alzheimer's disease and breast cancer, that develop later in life. Those people at greatest risk can then be targeted, screened at regular intervals and given appropriate advice about how to reduce the risk.

→ to eliminate all risk of the disease by correcting or replacing the faulty allele.

What are the advantages and disadvantages of gene testing?

Gene testing has already dramatically improved lives. Tests may be used to:

→ clarify a diagnosis and direct a physician toward appropriate treatments.

→ allow families to avoid having children with devastating diseases or identify people at high risk for conditions that may be preventable

However, there are concerns about gene testing.

→ For adult-onset disorders, such as Alzheimer's disease and some cancers, the tests are targeted at healthy people identified as being at high risk because of a strong family medical history for the disorder. A serious limitation of these susceptibility tests is the difficulty in interpreting a positive result because some people who carry a disease-associated mutation never develop the disease.

→ A limitation of all medical testing is the possibility of laboratory errors such as sample misidentification and contamination of the chemicals.

→ Other concerns are uncertainties surrounding test interpretation, the current lack of available medical options for these diseases, the test's potential for provoking anxiety, and risks for discrimination and social stigmatization.

Checkpoint 1

What is a genetic probe and what is its function?

Checkpoint 2

What is meant by the term 'presymptomatic testing' for predicting adult-onset disorders such as Huntington's disease?

Grade booster

What are your views on the development of a national DNA database?

Examiner's secrets

The identification, sequencing and mapping of genes by the Human Genome Project is central to medical advances in genetic research and is likely to be tested in an exam.

Exam practice answers: page 187

1. The Human Genome Project was completed in 2003 and is likely to be helpful to genetic engineers in the future. Describe the aims of the project and how the information provided will make it possible to eliminate all risk of genetic disease.

Genetic counselling and gene therapy

The aim of gene therapy is to treat a genetic disease by replacing defective genes in the patient's body with copies of a new DNA sequence. The Human Genome Project has enabled scientists to identify precisely many more genes responsible for genetic disease. The main problem involves developing a gene delivery system, that is, a means of inserting 'normal' versions of genes into a person's cells and ensuring that they function correctly once they get there. Genetic screening has raised a number of ethical and legal issues.

Gene therapy

Gene therapy usually requires a vector to introduce the DNA. Most procedures use viruses as vectors to deliver the selected gene to the target cells. Some use liposomes and others use injection of naked plasma DNA. There are two possible ways of replacing defective genes:

→ Gene therapy, involving somatic cell therapy targets cells in the affected tissues. The use of stem cells, rather than mature somatic cells, is longer-lasting in patients.
→ Germ-line therapy involves the introduction of corrective genes into germ-line cells, that is, the gene is replaced in the egg and will enable genetic corrections to be inherited.

Cystic fibrosis

Cystic fibrosis is due to a defective autosomal recessive allele, and to inherit the disease both parents must be carriers of these alleles. In the UK one person in 2,000 suffers from the condition. Sufferers produce a thick, sticky mucus from the epithelial cells lining certain passageways in the body. These secretions lead to a number of problems:

→ The pancreatic duct becomes blocked preventing enzymes from reaching the duodenum and so food digestion is incomplete
→ The bronchioles and alveoli of the lungs become clogged, causing congestion and difficulty in breathing. The mucus is difficult to remove and leads to recurrent infections

The normal gene codes for the production of a protein are found in the cell membrane. This protein, called cystic fibrosis transmembrane regulator (CFTR), transports chloride ions out of cells into mucus. Sodium ions follow out of the cells and water passes out of the cell by osmosis. This makes the mucus that lines the air passages a watery consistency. The protein of cystic fibrosis sufferers lacks just one amino acid and so cannot perform its transport function. Microbiologists have succeeded in isolating and cloning the gene which codes for the CFTR protein. The gene therapy technique works as follows:

Gene amplification using the polymerase chain reaction (PCR)

At a crime scene a small sample of DNA may be found, perhaps a single strand of hair or a small spot of blood. To carry out numerous laboratory tests larger samples are required. Using the PCR technique copies of specific fragments of DNA may be made. PCR rapidly produces many billions of molecules from a single DNA molecule. This enables tests to be carried out accurately and more rapidly. PCR is really semi-conservative replication of DNA in a test-tube. The sample of DNA is dissolved in a buffer and mixed with the enzyme DNA polymerase, nucleotides and short pieces of DNA called primers (which act as signals to the DNA polymerase to start copying).

The jargon

Exons are regions of DNA that code for proteins. Regions of non-coding DNA are called *introns*.

Checkpoint 3

In forensic science what is the purpose of the polymerase chain reaction?

Grade booster

Research the issues raised by storing genetic data.

Exam practice answers: page 188

1 The diagram shows the results of a test which can be used to analyse evidence left at the scene of a crime. This can then be compared with samples taken from various suspects.

Scene of crime   Suspect 1   Suspect 2   Suspect 3   Suspect 4

(a) (i)   What is the name given to this technique?
   (ii) State 2 types of evidence left at the scene of the crime which may have been used to provide the DNA sample.
   (iii) Which suspect has been incriminated by the results and give a reason for your answer?
(b) Describe the part played by the following in the technique.
   (i)   Restriction endonuclease.
   (ii) Gel electrophoresis.
(c) Other than criminal investigations, give another situation where this technique is used.

Applications of reproduction and genetics

In the past, conventional breeding techniques have been used to improve farm animals, crop and ornamental plants. However, the selection and cross-breeding is laborious, time-consuming and sometimes unpredictable. Micropropagation provides a rapid method for obtaining large numbers of genetically identical plants. In the future the quality of farm animals may be improved by laboratory-based breeding techniques involving embryos.

Micropropagation

Checkpoint 1

List the differences between asexual and sexual reproduction.

Asexual reproduction involves only one organism and the individuals produced are genetically identical, i.e. belong to a clone. The technique of micropropagation is based on the ability of the differentiated plant cell to give rise to all the different cells of the adult plant, i.e. plant cells are **totipotent**. Micropropagation is sometimes referred to as a test-tube plant culture. It is an extremely cost-effective way of producing large numbers of genetically identical plants, which are clones of a single parent. Four steps are generally followed in the micropropagation and are shown in the diagram below.

The advantages of micropropagation are:
→ large numbers of plants can be grown in sterile controlled conditions ensuring a greater survival rate than would be the case if seeds were planted outside
→ uniformity of crop
→ cold storage of large numbers of plants in a small space with reduced heating and lighting costs
→ perpetuation of unique genotypes
→ reduced space required for transport
→ pathogen-free when planted out

The jargon

An *explant* is a small segment of stem or bud.
A *callus* is an undifferentiated mass of plant tissue.

Checkpoint 2

By which cell division process does an explant grow into a callus?

Checkpoint 3

Why is uniformity in a crop plant an advantage?

Action point

By referring to the diagram, write out a bullet list to describe the steps in micropropagation.

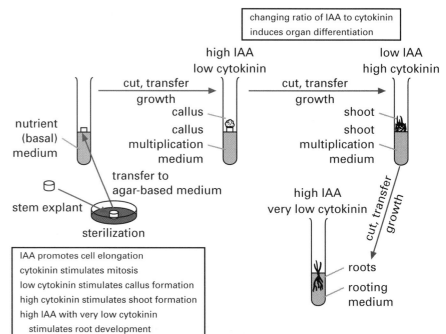

changing ratio of IAA to cytokinin induces organ differentiation

IAA promotes cell elongation
cytokinin stimulates mitosis
low cytokinin stimulates callus formation
high cytokinin stimulates shoot formation
high IAA with very low cytokinin
    stimulates root development

The disadvantages of micropropagation are:

→ requires aseptic conditions otherwise bacterial or fungal contamination of the culture medium results with subsequent loss of plants

→ the plants are genetically unstable, with an increased rate of mutation in medium-grown cells leading to abnormality in the plantlets. Regular inspection is needed to remove any defective individuals, thus labour costs are higher than with traditional propagation methods.

Cloning of animals

Embryo cloning

Eggs are taken from high milk yielding cows and are fertilised in a petri dish using sperm from the best bulls. This is known as *in vitro* fertilization (commonly called 'test-tube fertilisation'). The fertilized egg divides to form an embryo consisting of a ball of cells. When split into separate cells each will develop into a new embryo, genetically identical (clone) to the original. The embryos are then transplanted into other cows called surrogates. This technique of embryo surgery has enabled farmers to increase their stock and also to conserve rare breeds.

Cloning by nuclear transplants

This technique allows clones to be produced from one individual. It involves transplanting a nucleus from a somatic cell into an egg cell. The following describes the procedure:

Cells are taken from the tissues of the udder of a sheep (the donor) and cultured in a medium which stops division.

An unfertilized egg is removed from a different sheep (the recipient) and the nucleus is removed leaving an egg cell without a nucleus.

The donor and recipient cells are fused together and allowed to divide producing a ball of cells.

The developing embryo is implanted into the uterus of another sheep (the host or surrogate).

The lamb born is genetically identical to the original donor sheep.

Cloning is the only technique that will conserve unique features for future generations. Conventional cross-breeding reshuffles the genes with a possible loss of unique characteristics.

Links

The related topic of conservation of rare breeds is discussed on pages 202 and 203.

Action point

Discuss the ethical implications of genetic engineering.

Grade booster

Consider the advantages of cloning animals over conventional methods of artificial selection.

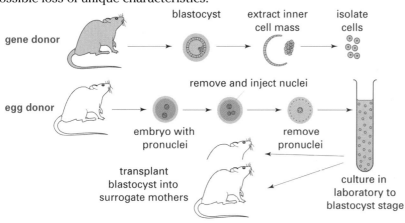

Action point

Describe the advantages and disadvantages of cloning animals.

Exam practice
answers: page 188

1 Define a clone and describe the principles involved in the cloning of plants and animals. Outline the advantages and disadvantages of this practice.

Tissue culture

The technique of growing cells in a laboratory is called tissue culture. The medium, in which the cells are grown, has to be precisely controlled and conditions such as water potential and temperature have to be carefully monitored. Animal cells in tissue culture develop into mature cells of the same type as the cell from which the culture was started. Cell cultures have been used for some time for medical and research purposes. New techniques such as cell replacement therapy and tissue engineering are being developed.

Checkpoint 1

Why are two fertilized eggs chosen to insert into the uterus?

Tissue culture

In theory, all cells can exist independently of the body. Cells from young animals and cancer cells can be induced to divide *in vitro*, provided they are supplied with nutrients. Most cells, however, differentiate into cells which have specific functions, such as nerve or muscle cells, and most of these specialized cells do not normally divide again.

Tissue engineering involves inducing living cells to grow on a framework of synthetic material to produce a tissue such as skin tissue, e.g. for the treatment of extensive, deep burns. Normally when the skin is damaged, stem cells in the epidermis divide many times to form new skin cells that grow around and cover the wound. If the burns are not too extensive surgeons may take skin from other parts of the body and graft it on to the wound. But when large areas are burned this may not be possible. In 1998 an artificial skin called 'Apligraf' was approved and is now widely used in place of skin grafts.

Checkpoint 2

Give examples of cells that retain their ability to divide in an adult.

Checkpoint 3

Give examples of how cell cultures have been used for medical and research purposes.

Stem cells

Other applications of tissue engineering include blood vessel replacement, bone and cartilage repair, and the treatment of degenerative nerve diseases. Central to this area of research is the use of stem cells. A stem cell is an undifferentiated cell capable of dividing to give rise to cells which can develop into different types of specialized cells. Stem cells have the ability to develop and form all the tissues of the body. The best sources of these cells are from very early embryos, but some adult tissues, such as bone marrow, also contain stem cells. Therapeutic stem cell cloning has enormous medical potential.

In many cases of traditional tissue transplantation, the patient's immune system rejects the donated cells. If the patient's own cells could be used to generate stem cells, then there would be no risk of rejection. Cloned stem cells could be used to generate organs for transplantation. This would prevent immune rejection and reduce the problem of organ shortages.

The ethics of using stem cells

There is considerable controversy surrounding this area of research. The supply of embryos comes from surplus embryos which were not placed into a female's uterus during fertility treatment. Once the stem cells are removed these embryos are then destroyed. Some think that it is not acceptable to use embryos for this purpose even if there was never any chance of the embryo being allowed to develop. Others consider that the potential benefits outweigh the ethical concerns. The human stem cells could be used to treat Parkinson's disease, Alzheimer's disease, heart disease, liver diseases, diabetes, multiple sclerosis and some cancers.

Opponents also argue that embryonic stem cell technologies are a slippery slope to reproductive cloning and can fundamentally devalue human life. That is, there is the possibility to clone humans. Those in the pro-life movement argue that a human embryo is a human life and is therefore entitled to protection. Supporters of embryonic stem cell research argue that such research should be pursued because the resultant treatments could have significant medical potential. It is also noted that excess embryos created for *in vitro* fertilization could be donated with consent and used for the research. The ensuing debate has prompted authorities around the world to seek regulatory frameworks and highlighted the fact that stem cell research represents a social and ethical challenge.

In vitro fertilization

IVF treatment involves mixing a female's egg with the partner's sperm in a dish where fertilisation takes place. Usually, ovulation is stimulated using hormones at a specific dosage which aims to cause several follicles to develop at the same time. The eggs are collected from the female and to each about 100,000 sperm are added. Two fertilised eggs are chosen and inserted into the uterus using a tube.

What happens to the 'unused' embryos has raised ethical issues. Some believe them to be human beings and should not be denied life. Others argue that females naturally lose embryos each month and that IVF embryos are no different. Others think that these embryos can be used as a source of stem cells, potentially a source of treatment for many genetic diseases. There is no world uniformity on this issue with some countries allowing this, while others have passed specific legislation to prevent it. One possible use of the embryos following IVF is to freeze them using liquid nitrogen, allowing them to be stored for many years. A female may have to undergo radiotherapy or chemotherapy as part of her treatment for cancer and this may damage her ovaries. This raises another ethical issue. Before implantation of the stored embryos can be carried out, permission of both partners is required. This can cause problems if the male partner no longer wants the woman to have his child.

Grade booster

Research the ethics of using stem cells. Discuss the ethical issues regarding what happens to the 'unused' embryos.

Examiner's secrets

Stem cell research is a topical issue that is likely to be tested in an exam.

Exam practice answers: page 188

1 Give terms to describe the following:
   (a) The technique of growing cells in a laboratory.
   (b) Replacing damaged cells such as those affected by skin burns.
   (c) A cell capable of dividing and forming new cells which can differentiate into specialized types of cells.
   (d) A group of genetically identical cells.

Genetically modified organisms

A genetically modified or transgenic organism is a plant, animal or microorganism that has its genotype altered to produce a new strain of organism. There is currently a great deal of opposition to GMOs and you should consider their benefits and possible problems associated with their use.

The jargon

A *transgenic* organism is one that has had the DNA from another organism of another species transferred to it.

Genetically modified organisms

Examples of GMOs are:

→ The production of insulin involving the introduction of human DNA into a bacterium.
→ A human gene that codes for an enzyme called antitrypsin, used to treat cystic fibrosis, has been transferred to sheep which now produce milk containing this enzyme.
→ The development of crop plants resistant to herbicides and pests, for example, by the transfer of genes that produce toxins with insecticidal properties from bacteria to higher plants such as potatoes.

Checkpoint 1

Give an example of a useful hormone produced by a genetically modified bacterium.

Transgenic plants

Bacteria are readily introduced into plants. Certain species naturally attack damaged plants and cause the plant cells to multiply and form a tumour. The bacteria do this by inserting genes from their own plasmids into one of the plant's chromosomes. The plasmid gene links with the DNA of the plant but stimulates the growth of a tumour. Scientists can replace the tumour-forming genes in the bacterial plasmids with useful genes.

The following are examples of GM crops:

→ Soya beans
   In many countries soya beans are very important as a source of food. They are used as an ingredient in a wide range of foods such as flour, protein and oil. About 60% of food products such as bread, biscuits, baby foods, soya milk etc. are soya based. Certain varieties of soya plants have been modified to be tolerant to a weed-killer. This allows the weed-killer to be sprayed onto the crop without affecting it and killing all the weeds. The weed-killer breaks down in the soil into harmless components.

→ Tomatoes
   Tomatoes ripen naturally as they produce an enzyme which breaks down the pectin in their cell walls. Tomatoes sold at the supermarket need to be firm and at their best when displayed. This creates a problem for supermarkets that need to transport tomatoes long distances from their supplier. Scientists have developed a genetically modified tomato called 'Flavr Savr'. A gene has been introduced into the tomato plant that effectively blocks the production of the enzyme. The result is that Flavr Savr tomatoes have a longer shelf life and a better taste.

Checkpoint 2

Is it possible that pollen may be transferred from GM crops to other plants?

Checkpoint 3

List advantages of GM crops.

Benefits and concerns

Are there potential risks in producing GM crops? Will the products and ingredients made from GM crops be labelled? Will the nutritional properties be affected? Will the environment be affected?

In theory on a world-wide scale GM crops have tremendous value:

→ **Solving food shortages** in various parts of the world; perhaps also enabling crops to be grown in drought areas.

→ **Producing improved food** with improved flavour and better keeping qualities.

→ **Reducing the harmful effects of modern farming** by:
  → Introducing nitrogen-fixing genes into crops such as rice and wheat, reducing the use of artificial fertilizers.
  → Introducing genes that confer resistance to insects, weeds and diseases.

Since the introduction of GM crops at the end of the 20th century, there has been an increase in public concern and some scepticism regarding the research carried out by government agencies. As a result many retailers have banned GM ingredients from their own brand products. This will remain until public confidence in the long term use of GM crops is restored.

Arguments against GM crops

→ Is the environment affected?
  There are concerns that the GM plants will transfer their genes to wild relatives or similar crops growing nearby with unforeseen effects. For example, plants with introduced genes that enable them to resist insect attack will quickly lead to the establishment of a resistant population of insect pests.

→ Is the food safe?
  GM crops contain marker genes. These are short, easily detected sequences of DNA used to determine whether the organisms have taken up the introduced gene. Some marker genes confer antibiotic resistance. There is concern that these genes may be transferred to the bacteria in the intestine of the consumer.

→ Will biodiversity be affected?
  Plant breeding may fall into the hands of just a few commercial companies and the varieties they offer to the farmer will be reduced. This could make crops more susceptible to attack by pests and diseases and lead to a reduction in the use of important old varieties of their wild relatives.

→ What will be the effect on organic farming?
  If GM crops are to be grown commercially in the UK, organic farm produce will be compromised. Pollen from GM crops, spread on the wind and by insects, could find its way into organic fields and beehives.

Grade booster

Research how GM soya beans modified to be tolerant to a weed-killer appear to have created many problems in countries such as Argentina.

Action point

Scientists have developed a genetically modified tomato called 'Flavr Savr'. Research how a gene has been introduced into the tomato plant that effectively blocks the production of the enzyme responsible for their ripening.

Exam practice   answers: page 188

1 What are the advantages and disadvantages of genetic engineering?

Answers
Reproduction and genetics

Checkpoints

1 Flowers brightly coloured, scented, nectar.
2 Sugar.
3 Large, with projections, and sticky.
4 Nutritive.
5 Three.
6 Eight.

Exam practice

1 Your answer should include 10 of the following points in a logical sequence:
- Anther produces pollen cells in four pollen sacs.
- Pollen sacs contain diploid spore mother cells each of which divides by meiosis to produce four pollen cells.
- Each pollen cell develops a thick outer wall or exine.
- Mitosis gives rise to a pollen tube nucleus and a generative nucleus.
- The carpel consists of stigma, style and ovary.
- The ovary contains an ovule(s) attached to the wall by a stalk (chalaza).
- The nucellus is surrounded by two integuments.
- The opening between the integuments is called the micropyle.
- The diploid ovule divides by meiosis to form four megaspores, three of which disintegrate.
- The fourth divides by mitosis to produce eight nuclei of the embryo sac.
- There are two polar nuclei in the centre of the cell.
- Three nuclei in antipodal cells at one end of the embryo sac and two synergid cells plus female gamete at the micropylar end.

Examiner's secrets

This question is in two parts so be careful to devote roughly equal amounts of time to describing the reproductive parts of the flower and to describing gametogenesis. Do not go into details of pollination and fertilization – the question does not ask for these.

Pollination and fertilization

Checkpoints

1 Self-pollination is the transfer of pollen to the stigma of the same flower whereas cross-pollination is the transfer to another flower on another plant of the same species.
2 Protandry – stamens ripen before stigma; monoecious – separate male and female flowers; incompatibility; heterostyli, etc.
3 To increase the chance of pollination and fertilization.
4 It gives a large surface area to filter pollen from the air and also for the easy release of pollen into the air.
5 Pollination is the transfer of pollen to the stigma of a

flower whereas fertilization is the fusion of gametes.
6 One of the male nuclei from the pollen tube fuses with both polar nuclei in the embryo sac to form a triploid nucleus. This divides mitotically.
7 Three.
8 It becomes the food store or endosperm.

Exam practice

1
- Pollen tube grows down the style.
- This is controlled by the tube nucleus.
- Followed by the generative nucleus.
- This divides mitotically to give two male nuclei.
- The pollen tube enters the embryo sac through the micropyle.
- One male nucleus fertilizes the female gamete to give the diploid zygote.
- The other male nucleus fuses with the two polar nuclei to form the triploid endosperm nucleus.

Seed development and germination

Checkpoints

1 They wither and die.
2 (a) The fruit wall or pericarp.
  (b) The seed.
3 So that energy is made available via respiration for metabolism and growth.
4 To produce substrates for respiration and the building blocks for synthesis.
5 For anchorage and water uptake.
6 Embryo.
7 Aleurone layer.

Exam practice

1 Your answer should include 10 of the following points in a logical sequence:
- The ovule develops into the seed.
- This forms an outer seed coat or testa.
- This encloses the endosperm and the embryo.
- The endosperm is the food reserve.
- It is formed by division of the triploid endosperm nucleus.
- The embryo develops from the fertilized egg cell.
- The embryo gives rise to the plumule and radicle and to the seed leaves or cotyledons.
- Germination starts with the uptake of water.
- This causes the seed to expand and the coat to rupture.
- It causes enzymes to become active.
- This mobilizes the food reserves in the cotyledons and/or endosperm.
- In cereals the aleurone layer of the endosperm makes the amylase for starch digestion.
- As a result of stimulation by the hormone gibberelin, the growing points (meristems) are reactivated.
- The radicle emerges first followed by the plumule.
- The importance of factors affecting germination.

This question is in two parts so ensure that equal time is devoted to seed development and to germination. You do not need to mention every answer in the mark scheme but a balanced answer will increase your chance of gaining maximum marks.

Human reproduction

Checkpoints

1 In the male the germinal epithelium of the seminiferous tubules is active from puberty throughout life; in the phase of multiplication few spermatocytes die; in the phase of maturation and meiosis each primary spermatocyte divides to form two secondary spermatocytes and then four spermatids which mature into sperms. In the female the germinal epithelium of the ovaries is active in the fetus only. In the phase of multiplication most oocytes die. In the phase of maturation and meiosis each primary oocyte divides to form a secondary oocyte and first polar body and further division is delayed until fertilization by a single sperm.
2 Mitosis.
3 32.
4 Eight.

Exam practice

1
- The male gonads are the sperm-producing testes.
- They are suspended outside the body in the scrotum in order to keep germinal tissue below core body temperature.
- Each testis contains numerous seminiferous tubules.
- These are surrounded by interstitial cells which secrete male hormones.
- The tubules unite to form the vasa efferentia.
- The sperm are stored in the epididymis.
- This leads to the vas deferens or sperm duct.
- This joins the urethra that leads from the bladder through the penis.
- Just before the sperm duct enters the urethra it receives the contents of the seminal vesicles.
- Surrounding the sperm duct/urethra junction is the prostate gland.
- Beyond the prostate the urethra receives the contents of Cowper's gland.
- One function of these glands from the following: secretion of sperm nutrients, activates sperm, secretion of enzymes.

Because the question asks for a diagram, marks are allocated accordingly. Your diagram should be accurate and well labelled and include an account of the functions of the parts you have labelled. However, make sure you do not waste valuable time by repeating yourself.

Hormonal control of reproduction

Checkpoints

1 After ovulation the concentrations of progesterone and oestrogen gradually increase, and then suddenly decrease. With the fall in level of these two hormones, the endometrium starts to disintegrate and menstruation starts. Oestrogen and progesterone regulate the production of FSH and LH by negative feedback. The high levels of oestrogen and progesterone inhibit the production of the gonadotrophic hormones from the pituitary, and the absence of LH in particular prevents ovulation. Keeping a high level of progesterone helps to maintain the uterus lining.
2 Day 14.
3 Testis.

Exam practice

1 (a) Cells of the follicle/theca; the development of the follicle mirrors the rise in oestrogen level.
  (b) Inhibits FSH production; fall in FSH with rise in oestrogen **or** promotes LH production; positive correlation between oestrogen and LH levels.
  (c) Maintenance of the lining of the endometrium.
  (d) While there is a high level of progesterone the pituitary does not produce FSH or LH; when the progesterone level falls the pituitary starts to secrete gonadotrophins again.
  (e) The pill contains progesterone, which inhibits FSH, preventing the development of ova.
2 (a) Oxytocin causes the contraction of smooth muscle of the myometrium.
  (b) Prolactin stimulates the secretion of milk; oxytocin causes the contraction of muscles to force the milk through the nipple.

Read the complete question before starting to write your answers. If you do not, you may include detail in one part which is requested in a later part of the question. You will then have to repeat yourself and so waste time. Try to deduce the answers from the information provided in the question rather than relying entirely on memory.

Fertilization and development

Checkpoints

1 To produce ATP to give energy for movement.
2 It prevents the entry of further sperm as only one is required to fertilize the ovum.
3 If the fetus is not the same blood group as the mother it will carry antigens foreign to the mother and she will produce antibodies which will pass to the embryo.
4 High levels of oestrogen and progesterone.

Exam practice

1 See text.

Meiosis

Checkpoints

1 A cell with a full set of chromosomes is diploid; in gamete production during meiosis this number is halved.
2 One member of the pair has come from the male and the other from the female.
3 Points where chromatids remain joined.
4 Random assortment of homologous chromosomes from the mother and father; recombination of segments of chromosomes during crossing over.

Exam practice

Your answer should include 10 of the following points in a logical order:

- During prophase 1 chromosomes shorten and thicken and so become visible.
- The nuclear membrane disappears.
- Homologous chromosomes form pairs or bivalents.
- They may entwine forming chiasmata.
- This is where DNA may break so that genetic material is exchanged, i.e. crossing over.
- During metaphase 1 each homologous pair lines up on the equator of the spindle.
- During anaphase 1 homologous chromosomes are separated as they move to opposite poles of the cell.
- This is caused by the contraction of the spindle fibres.
- The second division takes place with the new spindle forming at right angles to the first.
- During anaphase 2 the centromeres split as the chromatids move to opposite poles.
- During telophase 2 the chromatids begin to uncoil, disappear and the nuclear envelope reappears.
- Four haploid cells are separated as a result of cell plate formation.
- Bivalents are attached by centromeres.

Variation and evolution

Checkpoints

1 In continuous variation there is a gradation of one character from one extreme to another, whereas in discontinuous variation there is a limited number of distinct forms.
2 Discontinuous.
3 Crossing over; independent assortment; chromosomes contributed from each parent.

4 An environmental factor that confers greater chances of survival and reproduction on some individuals than on others in a population
5 Mutation that occurs in a cell other than a gamete-producing cell.

Exam practice

1 (a) Increases the range of weight and IQ but not head width and height. Weight and IQ are influenced by both genotype and environment, whereas head width and height are determined only by the genotype.
   (b) As a comparison with identical twins reared in the same family; acts as a control.
   Non-identical twins show a greater range of variation in their characteristics because they have fewer alleles in common.

Mendel's laws

Checkpoints

1 A gene controls a character or protein; an allele codes for an alternative form of the same character or protein and occupies the same locus.
2 Pollen is transferred to the stigma of the same flower.
3 The pea exhibits sharply contrasting characters and is relatively unaffected by environmental factors.
4 $608 \times \frac{1}{16} = 38$

Exam practice

1 You should indicate the correct genotype and phenotype for both parents using an actual example you have studied, e.g. yellow round YYRR × green wrinkled yyrr.
   Indicate the two types of gametes, i.e. YR and yr.
   Indicate the F1 phenotype and genotype, i.e. Yy Rr.
   Cross (a) should be Yy Rr × Yy Rr and should show four types of gametes (use a Punnett square).
   Give the phenotypes and genotypes of the offspring (F2) and the 9:3:3:1 ratio.
   Cross (b), which is in effect the back-cross, is Yy Rr × yy rr.

Deviations from Mendel's laws and sex linkage

Checkpoints

1 A ratio of 1 red (RR), 2 roan (RW), 1 white (WW).
2 1:1.
3 A carrier female and a colour-blind male.

Exam practice

1 (a) The recessive allele, which is carried on the X chromosome, is sex-linked. It occurs in small numbers in the gene pool, therefore there is little chance of a carrier and a haemophiliac reproducing. Affected females die at puberty or during childbirth.

(b)

Parental phenotype	Carrier mother		Haemophiliac father	
Parental genotype	XX^h		X^hY	
Gametes	X	X^h	X^h	Y
Possible offspring genotypes	X^hX	XY	X^hX^h	X^hY
Offspring phenotypes	carrier female	normal male	affected female	affected male

Mutations

Checkpoints

1 All mutations are expressed in prokaryotes but in diploid eukaryotes only the dominant mutations are expressed (due to the heterozygous condition).
2 Because they are recessive.
3 1 in 4, or 25%.
4 In parts of the world where malaria is endemic the condition causes mild anaemia but is an advantage because Plasmodium cannot enter abnormal red cells.
5 Because somatic cells are body cells other than the gametes.

Exam practice

1

- *T. monococcum* and **A. speltoides** are different species and produce sterile hybrid Z with 14 chromosomes.
- Their chromosomes differ in shape or size and so cannot pair during meiosis; so no gametes form.
- The sterile hybrid can only reproduce asexually.
- When a mitotic spindle fails to form by chance the chromosome number is doubled and a tetraploid form is produced.
- This can now form homologous pairs and can produce gametes or haploid cells during meiosis.

Human Genome Project

Checkpoints

1 A short piece of DNA the sequences of which are complementary to mutated sequences.
2 Huntington's disease is a rare mutation caused by a defective dominant allele. The disabling symptoms only become apparent when the patient is in his 30s or 40s. By this time the individual may have had children and so unintentionally passed the defective allele on to the next generation.

Exam practice

1 Your essay answer should include the following information:
Recording the complete base sequence of DNA; identifying all the genes formed by the bases; finding the location of the genes on the 23 human chromosomes; storing this information on a database.
Using a sample of DNA to identify if an individual is a carrier, or identify genes that play a contributory role in genetic disease. Correcting or replacing a faulty allele.

Genetic counselling and gene therapy

Checkpoints

1 1 in 4.
2 An artificially prepared sequence of DNA which can be used as a genetic 'marker'.
3 Chest physiotherapy and drugs to aid digestion.
4 Possible damage to the fetus.

Exam practice

1 The aim of gene therapy is to treat genetic disease by replacing defective genes in the patient's body with copies of the healthy gene. This could be carried out by replacing the gene in the egg, but this approach has not been taken because it could affect the human genome. Instead it may be possible to target somatic cells in the affected tissues.

With cystic fibrosis the approach is to wrap cloned healthy genes in microscopic lipid envelopes and at intervals to spray these on to the surface of the diseased lungs by means of an aerosol spray. It is hoped that the tiny liposome packages will be taken into the cells of the lungs' surface and the genes may enter the chromosomes and be expressed.

The question does not ask for the symptoms of cystic fibrosis. 'Treatment' could refer to physiotherapy and drugs, but in this context it applies to gene therapy. You should be able to provide a detailed up-to-date answer.

Genetic fingerprinting

Checkpoints

1 A genetic fingerprint is a person's DNA profile.
2 Subjecting chemicals to an electric current in a trough of gel.
3 PCR is to produce a large quantity of DNA from a small sample such as a smear of blood.

Exam practice

1 (a) (i) Genetic/DNA fingerprinting/DNA profiling.
     (ii) Blood/saliva/skin/hair/semen.
     (iii) Suspect 2. The only one which has the banding pattern which exactly matches that found at the scene of the crime.
   (b) (i) Used to cut the DNA into sections.
     (ii) Separation of fragments of DNA.
   (c) Paternity suits.

Applications of reproduction and genetics

Checkpoints

1 Asexual: no mixing of genetic material, no gametes, one parent, rapid, etc.
   Sexual is opposite of these.
2 Mitosis.
3 Ease of harvesting, all high yielding, etc.

Exam practice

1 Your answer should include the following points in a logical order:
- A clone is an asexually reproduced or genetically identical line of cells or organisms.
- A clone may be natural, e.g. bulbs, or may be produced in cell culture/micropropagation of plants.
- Cloning of plants involves the growth of a plant from a part of a plant or from a few cells.
- Cloning of animals involves the transfer of a nucleus from one individual to an egg from which the nucleus has been removed and the subsequent development of this egg.
- The embryo is placed in a host mammal surrogate, e.g. as in 'Dolly the sheep'.
- The embryo can be split before cell differentiation/ totipotency (plant/animal) therapeutic cloning of cells for repair and grafting.
   The advantages are as follows:
- Cell culture is useful in producing cancer cells for medical research, monoclonal antibodies.

- Produces a single identical genetic line of animal or plant with desirable characteristics.
- Maintains genetic stocks.
- In plants it may be much quicker than seed propagation.
   The disadvantages are as follows:
- In mammals and micropropagation in plants the technique is very expensive and can be unreliable.
- In plants disease or the entry of pathogens may cause problems.
- There may be objections on moral grounds.
- There may be inadvertent selection of disadvantageous alleles.
- There may be premature ageing.

This is a topical question but do not get carried away with a description of 'Dolly the sheep'. The question is more wide-ranging than this and asks for the principles of cloning in animals **and** plants and you should highlight the differences between them. The question also asks about the advantages and disadvantages so give an equal number of points for and against cloning. There is always a tendency to give details about the morality aspect of genetic engineering. It is a valid point but you must give a balanced answer.

Tissue culture

Checkpoints

1 To increase the chance that at least one will implant.
2 New cells to heal wounds, to replace skin cells, to produce new blood cells and to replace cells in the digestive system.
3 In the culture of viruses for vaccine production, and also in the production of monoclonal antibodies.

Exam practice

1 (a) Tissue culture.
   (b) Cell replacement therapy.
   (c) Stem cell.
   (d) Clone.

Genetically modified organisms

Checkpoints

1 Insulin.
2 The distances required to prevent pollen transfer between crops are the subject of controversy and disagreement, although such measures have been commonplace in conventional seed production for many decades. Transfer is likely to occur, but given stringent separation distances some claim that it should be possible to limit the range and degree of any spread.
3 Producing improved food with improved flavour and better keeping qualities. Reducing the harmful effects of modern farming by introducing nitrogen-fixing genes into crops such as rice and wheat, reducing the use of artificial fertilizers; introducing genes that confer resistence to insects, weeds and diseases.

Exam practice

1 Your answer should contain the following points. The advances made in genetic engineering have numerous advantages:

- The large-scale production of complex proteins or peptides that cannot be made by other methods.
- The production of higher-yielding crops with superior keeping qualities, etc.
- The health benefits for treating genetic diseases.

The new techniques are technically complicated and therefore very expensive on an industrial scale. There are concerns that the development of genetic engineering will introduce the possibilities of misuse. Germ-line gene therapy involves replacing the defective gene or allele with healthy genes inside the fertilised egg. This introduces the ethical question as to whether scientists have the right to alter the genotype of future generations. Do scientists know enough about how genes interact with each other? Some genes have no apparent function other than to control or 'switch off' other genes on the same chromosome. Tampering with genes in the fertilized egg could result in unforeseen effects in future generations.

There are a number of potential hazards associated with genetic engineering and it is impossible to predict what the consequences might be of releasing genetically engineered organisms into the environment. The potential hazards are:

- A new gene, on insertion, may disrupt normal gene function. For example, a potentially dangerous microorganism with a new gene may become a dangerous pathogen if it is released into the environment.
- Bacteria readily exchange genetic material. The recombinant DNA might get into other organisms. For example, herbicide resistance might be transferred to a weed species.
- The deliberate use of antibiotic-resistant genes in *E. coli* which lives in the human gut and these genes could be accidentally transferred to human pathogens.
- The possibility of transfer of DNA with linked pathogenic genes, for example, oncogenes increasing cancer risks.
- Scientists are confronted with vast quantities of data, a large proportion of which is genetic fingerprinting information. This raises concerns regarding the storage of this information as well as who has access to the data.

Scientists take precautions to minimize risk. Work on potentially dangerous pathogens has been restricted to isolated, highly specialized laboratories. However, despite legislation restricting the production and field testing of transgenic plants and animals there remain reservations about the long-term effects of the manipulation of the human genome and the production of GM organisms.

Revision checklist
Reproduction and genetics

By the end of this chapter you should be able to:

1	Interpret graphs showing the blood concentrations of FSH, LH, progesterone and oestrogen in the oestrous cycle.	Confident	Not confident **Revise** page 158
2	Describe with the aid of diagrams the process of meiosis and explain its significance.	Confident	Not confident **Revise** pages 162, 163
3	Discuss the genetic and environmental causes of variation.	Confident	Not confident **Revise** page 164
4	Define the term selection pressure and explain its importance in evolution.	Confident	Not confident **Revise** page 165
5	Use genetic diagrams to solve problems.	Confident	Not confident **Revise** page 169
6	Explain the causes and significance of gene and chromosome mutation.	Confident	Not confident **Revise** page 170
7	Understand the aims of the Human Genome Project and its value in the study of genetic disease.	Confident	Not confident **Revise** page 172
8	Consider the ethical and legal issues raised by genetic screening.	Confident	Not confident **Revise** page 175
9	Define the term genetic fingerprinting and explain the technique involved.	Confident	Not confident **Revise** page 176
10	Understand the meaning of the term cloning and its value in agriculture.	Confident	Not confident **Revise** page 179
11	Discuss the ethical issues regarding stem cell research.	Confident	Not confident **Revise** page 181
12	Consider the advantages and disadvantages of GM crops.	Confident	Not confident **Revise** pages 182, 183

Classification, biodiversity and evolution

It has been estimated that there are perhaps as many as 10 million kinds of organism living on the Earth today and an even greater number have become extinct. The huge diversity of life forms has come about due to the process of evolution. It is necessary to study the principles of modern classification and how biodiversity came about. You also need to appreciate the effect that human influences have had on evolution through selective breeding and how humans have contributed to the extinction of species.

Exam themes

- The principles of modern classification
- The important features of the five kingdoms
- The processes which affect allele frequencies in populations
- Examples of how environmental factors can act as stabilizing or evolutionary forces of natural selection
- The role of isolating mechanisms in the evolution of a new species
- The importance of genetic diversity
- The need to preserve genetic resources
- Biodiversity and extinction

Topic checklist

	Edexcel		AQA		OCR		WJEC	
	AS	A2	AS	A2	AS	A2	AS	A2
Principles of taxonomy	O		O		O		O	
The five-kingdom classification	O		O		O		O	
Population genetics				●		●		●
Evolution and selection	O		O		O			●
Isolation and speciation	O			●	O			●
Biodiversity, extinction and conservation	O	●	O		O		O	

Principles of taxonomy

It is believed that there are between three and 30 million species of living organisms on Earth but, to date, only about two million different kinds of organism have been described and identified. The sorting of living organisms into groups of a manageable size is known as taxonomy or classification. You should know about the principles of modern classification that show how organisms may be related through evolution by the number of characteristics they share.

Organisms are grouped in a hierarchical scheme

Taxonomy is the scientific study of the diversity of living organisms. A classification system places order and a general plan upon the immense diversity of life.

When describing living organisms taxonomists look for differences and similarities between them.

→ If organisms have evolved along the same lines similarities may occur, for example, the different mouth parts of insects. The study of the physical appearance of organisms allows biologists to group organisms together according to the extent to which they look alike.

→ It is also possible to study the extent to which organisms are related to one another. Organisms with common ancestors are classified close together and those more distant relatives are classified further apart. Here, biochemical features of protein and DNA composition are more important than physical resemblance.

In a system of classification, scientists place similar organisms closely together and dissimilar ones further apart. A classification system based on large groups being divided up into progressively smaller groups is said to be hierarchical.

The natural classification used today was devised by the Swedish scientist, Linnaeus, in the eighteenth century. In this scheme organisms are grouped together according to their basic similarities. A hierarchical system has been devised to distinguish large groups of organisms with a series of rank names to identify the different levels within the hierarchy. A **taxon** is a level in this classification hierarchy and is a collection of organisms sharing some basic features.

Starting with the smallest group:

→ **species** – a group of organisms which share a large number of characteristics and which can interbreed to produce fertile offspring, e.g. *Locusta* (locust).

→ **genus** – a group of species that are very closely related, e.g. *Locusta migratoria* (migratory locust).

→ **family** – a grouping of similar genera (plural of genus).

→ **order** – grouping of related families, e.g. Orthoptera (includes locusts, grasshoppers and crickets).

→ **class** – a grouping of similar orders, e.g. Insecta (insects).

→ **phylum** – a large grouping of all the classes that share some

Checkpoint 1

Write out the scheme of taxa (plural of taxon) within the classification system in order of size, starting with kingdom and ending with species.

Watch out!

Homologous should not be confused with analogous characters which are based on similarity in function.

Watch out!

The definition of a species has exceptions. The ability to breed is not the sole or essential diagnostic of a species.

Checkpoint 2

How can you justify classifying a corgi, an Alsatian and a Labrador as members of the same species?

Checkpoint 3

Why is it important for biologists to use Latin names for organisms mentioned in research articles?

features, e.g. Arthropoda (includes insects, spiders, centipedes and millipedes, crustaceans).
→ **kingdom** – the largest taxonomic grouping, e.g. animals, plants.

→ Moving up the hierarchy from species to kingdom means that the relationship between organisms in the groups decreases.
→ Moving down the hierarchy from kingdom to species means that the organisms are more closely related.

Binomial system

Many living organisms have common names, which may differ from one part of a country to another. This can be confusing and even more confusing if a particular species needs to be named and described in a scientific research paper, which may be read by a scientist from a different country.

To overcome the problem, organisms are named according to the **binomial system**. This system was introduced by Linnaeus in 1753 and is based on using Latin as an international language. Each organism is given two names, the name of its genus and the name of its species. This system means that an organism is given precise identification world wide, whereas the common name is not. The binomial system is still successfully used today because not only does each particular organism have its own unique scientific name but it also allows biologists to recognise that two species are closely related, e.g. *Panthera leo* (lion) and *Panthera tigris* (tiger).

Examiner's secrets

This is an example only. You are not expected to memorize the taxa or the features.

Checkpoint 4

As one progresses up the hierarchy do the similarities between organisms increase or decrease?

Action point

Research the classification of a chosen organism.

Don't forget

Biochemical homology, established by techniques such as amino acid sequence analysis and DNA hybridization studies, provides further evidence for evolutionary affinities within certain groups.

Grade booster

Research the recent approaches to using DNA analysis to influence classification.

Exam practice   answers: page 204

1 The two-spot ladybird, *Adalia bipunctata*, is a common British beetle. Complete the table below that classifies *Adalia bipunctata*.

Kingdom	
Phylum	
	Arthropoda
	Insecta
	Coleoptera
	Coccinellidae
Genus	
Species	

2 Discuss how DNA analysis has been useful in studying evolutionary relationships.

The five-kingdom classification

Until recently it was common to divide all living organisms into two kingdoms, plants and animals. Because there are some organisms which can be classified in both kingdoms and some which do not fit into either, the five-kingdom classification was proposed. You need to know about the distinguishing features of the five kingdoms.

Living organisms are divided into five kingdoms: Prokaryotae, Protoctista, Fungi, Plantae and Animalia.

Prokaryotae

The Prokaryotae are unicellular organisms and comprise:

→ bacteria
→ blue-green or cyanobacteria

They have no internal cell membranes, no nuclear membrane, no endoplasmic reticulum, no mitochondria and no Golgi apparatus. They possess a cell wall but it is not made of cellulose.

Protoctista

The members of this kingdom are mostly small unicellular or multicellular eukaryotic organisms, with membrane-bound organelles and a nucleus with a nuclear membrane. In this kingdom are found the organisms that are neither plants nor animals nor fungi, in fact all the organisms that do not fit elsewhere! So the kingdom includes algae, the water moulds, the slime moulds and the protozoa.

Fungi

The members of this kingdom are eukaryotic, the body consisting of a network of threads called hyphae, forming a mycelium. There is a rigid cell wall made of chitin. There are no photosynthetic pigments present and feeding is heterotrophic; all members of the group are saprophytic or parasitic. In some subgroups, the hyphae have no cross-walls, but in others cross-walls, or septa, are present. Reproduction is by spores which lack flagella. Examples are *Penicillium*, yeast, mushroom.

Plantae

The members of this kingdom are multicellular and photosynthetic. The cells are eukaryotic, have cellulose walls, vacuoles containing cell sap and chloroplasts containing photosynthetic pigments.

Some of the phyla of the Plantae include:

→ Bryophyta – mosses and liverworts
→ Filicinophyta – ferns
→ Coniferophyta – the conifers
→ Angiospermophyta – the flowering plants

Watch out!

Be aware that there are several different approaches to classification, e.g. cladistics and orthodox taxonomy.

Checkpoint 1

Why is the Protoctista sometimes referred to as 'the ragbag kingdom'?

Checkpoint 2

Some fungi are plant-like in appearance. Why are they now placed in a separate kingdom?

Checkpoint 3

Which group of organisms *could* form a sixth kingdom?

The flowering plants are the predominant plant group. They include all our major food plants. Their flowers have seeds which are enclosed in a fruit formed from the ovary wall. Angiosperms are divided into two classes.

→ Monocotyledons – embryo has one seed leaf. The group includes all the grasses, which includes cereals.
→ Dicotyledons – embryo has two seed leaves. The group includes trees such as oak and beech, shrubs and flowers.

Animalia

The members of this kingdom are multicellular, heterotrophic, eukaryotes, cells lacking a cell wall, exhibit nervous coordination. The animal kingdom is divided into 20 major phyla and several minor ones. They include:

→ Non-chordates (invertebrates)
→ Cnidaria (jellyfish)
→ Platyhelminthes (flatworms)
→ Nematoda (roundworms)
→ Annelida (segmented worms)
→ Mollusca (molluscs)
→ Arthropoda (arthropods)
→ Echinoderms (starfish)
→ Chordata (vertebrates)

Grade booster

You do not need to learn the names of the animal phyla.

Checkpoint 4

To which phylum do humans belong?

Most of the chordates have a vertebral column and are therefore referred to as vertebrates. The vertebrate chordates are divided into the following classes:

→ fish
→ amphibians
→ reptiles
→ birds
→ mammals

95% of all animals are invertebrates.

Exam practice   answers: page 204

1 The diagram below classifies living organisms into five kingdoms.

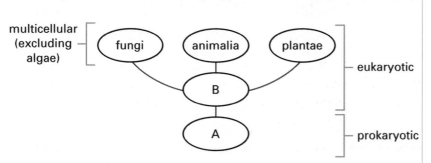

(a) Name kingdoms A and B.
(b) Consider the three kingdoms Fungi, Animalia and Plantae. Give one feature of each that distinguishes it from the other two.

Population genetics

Population genetics is concerned with the factors that determine the frequencies of alleles in populations of organisms. You will also need to understand the connection between allele frequency and phenotype frequency.

Gene pool

A population of organisms reproducing sexually contains a large amount of genetic variation called a **gene pool**. Each organism contains just one of the many possible sets of genes that can be formed from the pool. The gene pool remains stable if the environment is stable. However, if the environment changes some phenotypes will be advantageous and will be selected for, while others will be at a disadvantage and will be selected against. Thus a gene pool is constantly changing, some alleles becoming more frequent and others less frequent. In some circumstances alleles may be totally lost from the gene pool.

Checkpoint 1

Define gene pool.

Genetic changes in populations

In large populations, assuming there is no selection, the proportion of dominant and recessive alleles of a particular gene remains constant. It is not altered by interbreeding. The Hardy–Weinberg principle is used to calculate allele and genotype frequencies in a population. It can therefore be used to predict the number of defective individuals in a population. The Hardy–Weinberg equation is

$$p^2 + 2pq + q^2 = 1$$

Consider a pair of alleles Aa; let p = frequency of A and q = frequency of a; in the population $p + q = 1$, assuming random mating, then in the next generation AA = p^2; Aa = $2pq$; aa = q^2.

> *Example*: a recessive allele confers resistance to an insecticide in a particular insect species. Explain how the allele is distributed if 36% of the insect population is resistant.
> rr = 36% or 0.36, i.e. $q^2 = 0.36$, $q = \sqrt{0.36} = 0.6$
> Since $p + q = 1$, $p = 0.4$, $p^2 = 0.16$
> $2pq = 2 \times 0.4 \times 0.6 = 0.48$ or 48%
> Therefore allele distribution is as follows:
> RR (p^2) = 0.16 (16%); Rr ($2pq$) = 0.48 (48%); rr (q^2) = 0.36 (36%)

Don't forget

It is not possible to observe whether organisms are homozygous dominant or heterozygous for a particular characteristic. They all look the same. The Hardy–Weinberg equation allows us to calculate the proportions of the alleles of a particular gene.

Sometimes, variations in gene frequencies in populations occur by chance. This is known as random **genetic drift**. It may be an important evolutionary mechanism in small or isolated populations. Say an allele occurs in 1% of the members of a species. In a large population, say 1 000 000, then 10 000 individuals may be expected to possess the allele. By chance, the population of individuals with the allele will not be significantly altered in the next generation. If, however, the population is much smaller, say 1 000 individuals, only one will carry the allele. By chance, this one may fail to mate and pass on the allele and so it will be lost from the population altogether.

An important case of genetic drift is when a few individuals become isolated from the rest of the species and start a new population, e.g. when a few individuals colonise an isolated island or some new habitat.

Checkpoint 2

Define genetic drift.

Checkpoint 3

Suggest factors that can affect allele frequencies over time.

These founder members of the new population are a small sample of the population from which they came. By chance, they may have a very different gene frequency. While the founder population remains small it may undergo genetic drift and become even more different from the large parental population. This process is called the **founder effect**. The effect undoubtedly contributed to the evolutionary divergence of Darwin's finches after strays from the South American mainland reached the remote Galapagos Islands.

Disasters such as earthquakes, floods and fires may reduce the size of a population drastically. The result is that the genetic makeup of the surviving population is unlikely to be representative of the makeup of the original population. By chance, certain alleles will be over-represented among survivors, other alleles will be under-represented and some alleles will be eliminated completely. Genetic drift may continue to affect the population for many generations, until the population is again large enough for sampling errors to be insignificant. Because alleles for at least some loci are likely to be lost from the gene pool, the overall genetic variability in the population is usually reduced.

Exam practice

answers: page 204

1 Two different varieties of the two-spot ladybird exist, the normal red form and a black form. These two varieties are determined by a single gene with the allele for black body, B, being dominant to that for red body, b. The bar chart below shows the frequency of the allele for black body in autumn and spring samples over a number of years.

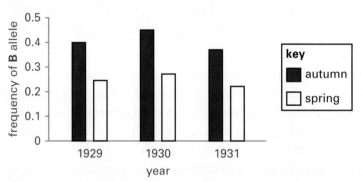

(a) (i) In the autumn 1929 sample, what was the frequency of the B allele and the b allele?

(ii) Use the Hardy–Weinberg expression to calculate the expected frequency in the autumn 1929 sample of ladybirds with the genotype Bb and black ladybirds.

(b) Some information about the biology of the two-spot ladybird is given below.

→ The two-spot ladybird spends the winter as an adult. When the weather gets warmer it becomes active and starts to feed and breed.

→ The temperature of the ladybird depends on that of its environment. The higher the temperature, the more active the ladybird.

→ Temperature sensors placed under the wing cases showed that the black areas cool down and heat up faster than the red ones.

Use this information to explain the seasonal changes in frequency of the B allele shown in the bar chart.

Examiner's secrets

The principle is based on two equations:
$p + q = 1$ (gene pool)
$p^2 + 2pq + q^2 = 1$ (total population).
If you are provided with suitable data always use the $p + q = 1$ equation first to calculate the frequency of alleles. From the figure obtained you can calculate the other parts of the equation.

Examiner's secrets

You may not be required to use the Hardy–Weinberg principle in calculations; check your specification.

Evolution and selection

Check the net

The modern reinterpretation of Darwin's theory is called neo-Darwinism (it is the restatement of the concept of evolution by natural selection in terms of Mendelian and modern genetics) and may be found at
www.panspermia.org/neodazrw.htm
homepages.which.net/~gk.sherman/baaaaabc.htm

Checkpoint 1

Explain why populations remain constant despite the production of large numbers of offspring.

The jargon

Adaptive radiation is the development from a single ancestral group of a variety of forms adapted to various environments.

Action point

Make your own notes on Darwin's finches and how their beaks are adapted to different feeding niches.

The jargon

Selection pressure is the term given to a factor that has a direct effect on the numbers of individuals in a population of organisms.

Evolution is the process by which new species are formed from pre-existing ones over a period of time. The basis of contemporary thought surrounding the theory of evolution was first put forward by Alfred Wallace and Charles Darwin. In 1859 Darwin proposed natural selection as the force that causes changes in populations. More recently biologists have realized that natural selection can also maintain variation and therefore stabilize a population.

Natural selection

Darwin's observations of variation within a population and the tendency for the adult population to be stable in size led to the development of the idea of natural selection. The theory is based on the following observations:

→ variation is found in all populations
→ individuals within a population have the potential to produce large numbers of offspring yet the number of adults tends to stay the same from one generation to the next

From these observations, two deductions were made.

→ There is a struggle for survival with only the 'fittest' surviving. The individuals that survive and reproduce pass on to their offspring the characteristics that enable them to succeed.
→ In time, a group of individuals that once belonged to the same species may give rise to two different groups that are sufficiently distinct to belong to two separate species.

During his voyage on HMS *Beagle*, Darwin found evidence of adaptive radiation in the finch population. He observed 14 species, differing greatly but particularly in beak morphology, reflecting differences in feeding habits. It was suggested that soon after the Galapagos Islands formed they were populated by a flock of finches from the mainland. These then evolved by natural selection to fill the available vacant niches, as other perhaps better-adapted forms that would have been in competition with the finches had failed to reach the islands.

When environmental conditions change, there is a selection pressure on a species causing it to adapt to the new conditions. This pressure determines the spread of an allele within the gene pool. Darwin proposed natural selection as the force that causes changes within populations. More recently biologists have realized that natural selection can also maintain variation if there is no need to change and therefore stabilize a population. The types of selection are as follows.

→ *Directional selection* is when an environment favours all the individuals carrying one gene at the expense of the individuals carrying an alternative.

→ *Stabilizing selection* results in several different types of individuals being maintained in a population so that genetic variation is maintained.

→ *Disruptive selection* favours both extremes of variants and leads to a divergence of the phenotype, i.e. to the emergence of two distinct forms. This form of selection is rare but is more likely to result in the formation of a new species.

Polymorphism

Polymorphism is the term used to describe the presence of clear-cut, genetically determined differences between large groups in the same population. Where a population is split into two subpopulations it is the result of conditions that favour more than one phenotype within the population. If gene flow between the subpopulations is prevented, then ultimately each population may give rise to a new species. Examples of polymorphism are colour and banding patterns in the land snail *Cepaea nemoralis* and ABO blood grouping in humans.

Checkpoint 2

Is variance increased or reduced by directional selection?

Action point

Draw normal distribution graphs to show the different types of selection.

Checkpoint 3

How may gene flow between populations be prevented?

Exam practice   answers: page 204

1 Shell colour and pattern in *Cepaea nemoralis*, the land snail, is genetically determined. Shells may be brown, green or pink. In addition, the whole shell may have up to five dark bands on it. Thrushes eat these snails, which they select by sight.

The table below shows the distribution of some shell types in three different habitats.

Shell type	Percentage of each shell type present in:		
	Grassland (green pasture)	Rough herbage (nettles, long grass and dead stems)	Woodland floor (dead leaves, soil and twigs)
Brown	10	23	54
Green	58	19	12
Pink	7	8	4
Banded	25	50	30

(a) Name the shell type with the greatest survival value on the woodland floor.

(b) Explain the relative abundance in grassland and rough herbage of snails with green shells and pink shells.

(c) Explain the relative abundance of banded snails in grassland, rough herbage and woodland.

(d) In winter the snails hibernate just beneath the soil with the mouth of the shell upwards and covered with white mucus. The thrushes dig out such prey from the soil. Suggest why you would not expect selection to operate on shell colour in these circumstances.

Isolation and speciation

In theory, any individual in a population is capable of breeding with any other. However, breeding subunits may become separated in some way, i.e. isolated, and evolve along separate lines. If, being reunited after many generations, individuals were found to be incapable of breeding successfully with each other, they would have become a separate species.

The origins of species

Within a population of one species there are groups of interbreeding individuals. Each of these breeding subunits is called a **deme**. New species arise when some barrier to reproduction occurs so that the gene pool is divided and different combinations of mutants develop in each pool division. Such a barrier which effectively prevents gene exchange is called an isolating mechanism. If the separation is long-term, eventually the two groups will be so different that two new species incapable of interbreeding are formed. This process is called speciation. For new species to develop from a population, some form of isolating mechanism is required.

There are two main types of isolating mechanism:

→ *Reproductive isolation within a population* occurs when organisms inhabiting the same area become reproductively isolated into two groups for reasons other than geographical barriers. Species formation occurring in demes in the same geographic area is known as **sympatric speciation**. The barriers to breeding include the following mechanisms:

  → Behavioural isolation – in animals with elaborate courtship behaviour the steps in the display of one subspecies fails to attract the necessary response in a potential partner of another subspecies.

  → Mechanical isolation – the genitalia of the two groups may be incompatible.

  → Gametic isolation – in flowering plants pollination may be prevented because the pollen grains fail to germinate on the stigma whereas in animals sperms may fail to survive in the oviduct of the partner.

  → Hybrid inviability – despite fertilization taking place development of the embryo may not occur.

  → Hybrid sterility (polyploidy).

→ *Geographical isolation between populations* occurs when the population becomes geographically split into separate demes; the evolution of a new species is very probable, given time. This sort of speciation is known as **allopatric speciation**.

Consider the isolation model shown in the diagram at the top of the opposite page. A population of birds with short flight range feed and breed only in the cool conditions of a valley and the lower slopes of two mountains a considerable distance apart.

Checkpoint 1

What are the forces for genetic changes in populations that have given rise to new species?

Examiner's secrets

You should understand how separation by geographical features, habitat changes, changes in body form and changes in breeding mechanisms may lead to the formation of a new species.

Checkpoint 2

Define speciation.

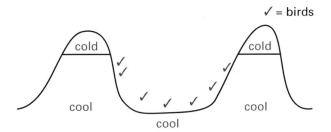

The climate then changes and it gets warmer (see diagram below).

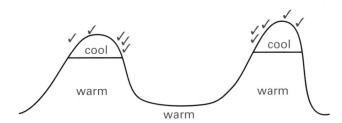

The birds are now confined to the cool mountain tops and the two populations are reproductively isolated. Given a very prolonged period of isolation, genetic drift and directional selection occurs, making the two populations genetically distinct.

If the climate reverts to the original situation shown at the top of the page, and the birds are again able to inhabit the valley and lower mountain slopes and the two populations can no longer interbreed, then they have formed two separate species.

Darwin considered that species gradually change over long periods of time from one form to another. It would then be expected that biologists would find intermediate forms between one fossil species and the next in successive rock strata. However, these forms are surprisingly rare and this has led some biologists to believe that new species may arise relatively rapidly (perhaps within a few thousand years) and then remain unchanged for millions of years before changing again.

Grade booster

In order to answer any practice on evolution a wide knowledge of the topic is needed. This can only be gained by background reading so that you will be able to quote actual examples. You should not only define the species but also discuss the difficulties posed in giving a single definition. You should then quote a specific example of isolation or use a 'model' you have studied. Geographical isolation of populations and the effects of natural selection on their gene pools should be the main theme of your answer.

Hybrid sterility

One particular form of reproductive isolation is hybrid sterility. A hybrid may be defined as the offspring resulting from cross-breeding of different species. When individuals of different species breed, the sets of chromosomes from each parent are different. These sets are unable to pair up during meiosis and so the offspring are unable to produce gametes. The hybrid is therefore sterile and the species is reproductively isolated. An example of hybrid sterility is a 'zebronkey'. This is the name given to the offspring which results from the mating of a zebra with a donkey. The zebronkey is sterile as it has 53 pairs of chromosones that are unable to form homologous pairs at prophase 1 of meiosis.

Exam practice
answer: page 205

1 Discuss the part played by geographical isolation in the formation of a new species.

Biodiversity, extinction and conservation

Biodiversity is a measure of the number of species on the planet; extinction is the loss of species. Over the last 200 years certain human activities have had negative effects on the environment and this has, in turn, affected the survival of plants and animals. Now scientists have come to realize that there is a biodiversity crisis, a rapid decrease in the variety of life on Earth.

Check the net

You'll find up-to-date information on biodiversity at www.greenpeace.org

Biodiversity and extinction

Human activities are altering ecosystems upon which they and other species depend. In the oceans stocks of many fishes are being depleted by over-harvesting, and some of the most productive and diverse areas, such as coral reefs and estuaries, are being severely stressed. Globally, the rate of species loss may be as much as 50 times higher than at any time in the past 100 000 years.

Extinction is a natural process that has been taking place since life first evolved. It is the current **rate** of extinction that underlies the biodiversity crisis. Scientists believe that the normal 'background' rate of extinction is one out of every million species per year. It is now estimated that human activity in tropical areas alone has increased extinction rates between 1 000 and 10 000 times! Massive destruction of habitats throughout the world has been brought about by agriculture, urban development, forestry, mining and environmental pollution. Marine life has also been affected. About one third of the planet's marine fish species rely on coral reefs. At the current rate of destruction about half of the reefs could be lost in the next 20 years.

Grade booster

Research the factors that led to the recent extinction of one named animal species and one named plant species.

Checkpoint 1

Define biodiversity.

Checkpoint 2

Define extinction.

Checkpoint 3

Name an ecosystem currently under threat due to human activity.

Checkpoint 4

State 3 main causes for the decline in numbers of large mammals such as gorillas and tigers.

Checkpoint 5

What is meant by species conservation?

Endangered species

The vast majority of Earth's earlier occupants, including the large and once dominant dinosaurs and tree ferns, have become extinct largely as a result of climatic, geological and biotic changes. At the present time, human activity has taken over as the main cause of species extinction. Many of the larger mammals such as mountain gorillas, giant pandas, tigers and polar bears are threatened.

Their decline in numbers has three main causes: loss of habitat, over-hunting by humans and competition from introduced species.

Other species are also threatened by additional causes such as deforestation, pollution and the drainage of wetlands.

It is now recognised that each species may represent an important human asset, a potential source of food, useful chemicals, or disease-resistant genes. For example, of the many plants growing in the tropical rain forests, there may be some with medicinal properties. The extinction of any plant species before their chemical properties have been investigated could amount to an incalculable loss. There is therefore a need for **species conservation**, the planned preservation of wildlife.

Conservation of genetic sources

Present-day plants and animals used in agriculture and horticulture have been developed from plants and animals that were originally in the wild.

Breeding increases genetic uniformity with the loss of rarer alleles. In the past breeders may have neglected some important qualities, such as resistance to cold and disease, etc. These need to be added back into highly cultivated varieties, using the wild plants and animals as a gene bank. If habitats, and the wildlife that live in them, are threatened this may no longer be possible.

In recent years there has been much concern about the loss of gene pools and legislation has been introduced in an effort to prevent the extinction of endangered species.

The following are some of the steps that have been taken:

→ Stocks of seeds of 'traditional' varieties of plants are stored in seed banks, and sperm banks have been established.
→ The founding of rare breeds societies to maintain old, less commercial varieties of animals.
→ The protection and breeding of endangered species in specialized zoos.
→ Reintroduction programmes, e.g. red kite in mid-Wales.
→ Global organisations, such as the World-Wide Fund for Nature, mount continuing campaigns to promote public awareness.
→ International cooperation restricting trade, e.g. in ivory and whaling.
　→ In the UK, the Nature Conservancy Council is the government body that promotes nature conservation and gives advice to government and to all those whose activities affect wildlife and their habitats.
　→ It produces a range of publications and educational material.
　→ It proposes schemes of management for each of the major ecosystem types, endeavouring to conserve species diversity.
→ It establishes nature reserves managed by wardens.

Legislation has been introduced to protect endangered species and to prevent over-grazing, over-fishing, hunting of game, collecting of birds' eggs, picking of wild flowers and collecting of plants.

Examiner's secrets

When presented with appropriate information, you will need to make balanced judgements between the need to meet the demands for increasing food production and the need to conserve the environment.

The jargon

Conservation is a dynamic process involving management and reclamation.

Watch out!

Don't confuse conservation and preservation.

Examiner's secrets

You should be prepared to discuss the economic and ethical reasons for conservation.

Check the net

You'll find excellent information about conservation issues at www.panda.org

Exam practice   answers: page 205

1 What are 'endangered species'? For what reasons have some species become endangered? How may species be conserved for future generations?

Answers
Classification, biodiversity and evolution

Principles of taxonomy

Checkpoints

1 Kingdom, phylum, class, order, family, genus, species.
2 Because they can interbreed to produce fertile offspring.
3 The binomial system uses Latin as an international language so that all biologists are presented with precise identification worldwide.
4 Decrease.

Exam practice

1 | Kingdom | Animalia |
Phylum	Arthropoda
Class	Insecta
Order	Coleoptera
Family	Coccinellidae
Genus	*Adalia*
Species	*bipunctata*

2 In recent years protein and DNA analysis has been used to confirm evolutionary relationships.

• The sequence of amino acids has been analysed in part of the fibrinogen molecule of various mammals. The sequence has been found to differ in varying degrees from one species to another and this has enabled scientists to draw up a possible evolutionary tree for mammals.

• On occasions it may be sufficient to use techniques, such as chromatography and electrophoresis, to separate and compare particular chemical substances present in different species. This may reveal similarities or differences which indicate how closely related the species are.

• It is possible to compare the sequence of bases in different organisms. The more alike the sequences, the closer the organisms are presumed to be in evolution. This technique is called DNA hybridisation.

• This comparison of DNA from a variety of living species can give a quantitative indication of how closely the genes are related. To work out how closely related are two species of primates, such as humans and chimpanzees, the DNA strands from both species are extracted, separated and cut into fragments. The fragments from the two species are then mixed and analysed. This technique gives results which show that chimpanzees and humans have 97.6% DNA in common, whereas humans and rhesus monkeys have 91.1% of DNA in common. Recent studies using this technique have also shown that the hippopotamus and whale are closely related.

The five-kingdom classification

Checkpoints

1 It contains all the organisms that cannot be fitted into any of the other kingdoms.
2 They do not possess chlorophyll.

3 Viruses.
4 Chordata.

Exam practice

1 (a) A, Prokaryotae; B, Protoctista.
  (b) Fungi: lack of chlorophyll, hyphae present, cell walls of chitin.
      Animalia: heterotrophic, nervous coordination.
      Plantae: autotrophic, possession of photosynthetic pigments, e.g. chlorophyll.

Population genetics

Checkpoints

1 The total of all the alleles of all the genes in a population.
2 Variations in gene frequencies in gene populations that occur by chance.
3 Emigration, reproduction or immigration, genetic drift, mutations.

Exam practice

1 (a) (i)   B = 0.4; b = 0.6
      (ii) Bb = 0.48
            Black ladybirds 0.64; you should show that black ladybirds may be homozygous dominant, or heterozygous.
  (b) The frequency of black ladybirds is higher in the autumn; black ladybirds absorb heat quicker and become active sooner; they therefore breed more; B allele increases.

Evolution and selection

Checkpoints

1 Predation, competition, etc.
2 Reduced.
3 Geographical and/or reproductive isolation.

Exam practice

1 (a) Brown.
  (b) Green shells are better camouflaged and are less conspicuous in green pasture than against a mixture of green and brown stems or a foliage of rough herbage. Pink shells are visible and readily selected in both habitats.
  (c) The banded pattern makes the shell shape less obvious. Rough herbage presents a mottled background on which banded shells are well camouflaged. Lighter bands show up against the uniform brown of the woodland floor.
  (d) The shell is concealed beneath the soil (its colour is hidden), so the thrush comes upon it by chance.

Isolation and speciation

Checkpoints

1 Mutations, crossing over during meiosis, independent assortment.
2 The process by which a new species is formed.

Exam practice

Your answer should include all the following points in a logical sequence:

- A description of how the population may become separated.
- For example, when it migrates, due to continental drift, seed dispersal or a geographical feature or physical barrier such as mountain, desert, ocean.
- This separation prevents populations from interbreeding, i.e. prevents the flow of genes.
- The isolated group may possess only a small proportion of the total variation that existed in the original population/gene pool.
- Thus there are different frequencies of alleles.
- Genetic drift may result.
- If environmental selection pressures differ, this will result in differences of phenotype between the two populations.

- This will continue until the original population forms two separate or new species that cannot interbreed if they are brought together.
- Give a suitable example for this, e.g. behavioural, incompatibility, etc.
- Definition of a species – can interbreed to produce fertile offspring.

Biodiversity, extinction and conservation

Checkpoints

1 Biodiversity is a measure of the number of species on the planet.
2 Extinction is the loss of species.
3 Coral reefs, estuaries, tropical rain forests.
4 Loss of habitat, over-hunting by humans, competition from introduced species.
5 Species conservation is the planned preservation of wildlife.

Exam practice

Most of the answers to this practice can be found in the text. However, there is a need to give detailed examples.

Revision checklist
Classification, biodiversity and evolution

1	Define the terms species, taxonomy, biodiversity, extinction and evolution.	Confident	Not confident **Revise** pages 192, 198, 202, 203
2	Describe the classification of species into the taxonomic hierarchy of kingdom, phylum, class, order, family, genus and species.	Confident	Not confident **Revise** page 192
3	Outline the binomial system of naming living organisms.	Confident	Not confident **Revise** page 193
4	Describe the characteristic features of the five kingdoms – Prokaryotes, Proctista, Fungi, Plantae and Animalia.	Confident	Not confident **Revise** pages 194–195
5	Define the terms gene pool and genetic drift.	Confident	Not confident **Revise** page 196
6	Explain Darwin's theory of natural selection.	Confident	Not confident **Revise** page 198
7	Explain the role of isolating mechanisms in the evolution of new species.	Confident	Not confident **Revise** pages 200–201
8	Discuss how variation, adaptation and selection are important components of evolution.	Confident	Not confident **Revise** pages 196–197
9	Outline the reasons for the conservation of plant and animal species.	Confident	Not confident **Revise** pages 202–203
10	Discuss the consequences of human activity on the biodiversity of plants and animals.	Confident	Not confident **Revise** page 202
11	Discuss the value of legislation to conserve endangered species.	Confident	Not confident **Revise** page 203

Resources

This section is intended to help you develop your study skills for examination success. You will benefit if you try to develop skills from the beginning of your course. Modern AS and A-level exams are not just tests of your recall of textbooks and your notes. Examiners who set and mark the papers are guided by assessment objectives that include skills as well as knowledge.

Exam board specifications

In order to organize your notes and revision you will need a copy of your exam board's syllabus specification. You can obtain a copy by writing to the board or by downloading the syllabus from the board's website.

AQA (Assessment and Qualifications Alliance)
Publications Department, Stag Hill House, Guildford, Surrey GU2 5XJ – www.aqa.org.uk

CCEA (Northern Ireland Council for Curriculum, Examinations and Assessment)
Clarendon Dock, 29 Clarendon Road, Belfast, BT1 3BG – www.ccea.org.uk

EDEXCEL
Stewart House, 32 Russell Square, London WC1B 5DN – www.edexcel.org.uk

OCR (Oxford, Cambridge and Royal Society of Arts)
1 Hills Road, Cambridge CB2 1GG – www.ocr.org.uk

WJEC (Welsh Joint Education Committee)
245 Western Avenue, Cardiff CF5 2YX – www.wjec.co.uk

The AS/A-level
specifications

Action point

Check with your teacher how your practical skills will be assessed.

The jargon

Every A-level specification includes synoptic assessment at the end of the A2 course. Synoptic questions draw on the ideas and concepts of earlier units and thus link topics.
Synoptic assessment is dealt with in more detail on pages 218 to 224.

All A-level courses are in two parts, with a number of separate modules or units in each part. Students will start by studying the AS (Advanced Subsidiary). Those wishing to do so will go on to study the second part of the course, called A2. There is some element of choice as to which modules are studied but this may depend on the Awarding Body.

Advanced Subsidiary

The AS modules are compulsory and they cover the common core of the subject. AS is designed to provide an appropriate assessment of knowledge, understanding and skills expected of candidates who have completed the first half of a full A-level qualification. The level of demand of the AS exam is that expected of candidates half-way through a full A-level course of study, i.e. between GCSE and A-level.

AS may be used in one of two ways:

→ As a final qualification, allowing you to broaden your studies and to defer a decision about specialism.
→ As the first half (50%) of an A-level qualification which must be completed before an A-level award can be made.

Thus the AS specification establishes core principles on which an understanding of biology is based, and at the same time covers relevant topics in sufficient depth to form coherent modules for students who may not continue their study of biology to A-level.

A-level (AS + A2)

The A-level exam course is in two parts:

→ The AS is the first half of the course and constitutes 50% of the total award. It will normally comprise three teaching and learning modules.
→ The A2 is the second half of the course, the remaining 50% of the total award. It also normally comprises a further three teaching and learning modules which, in addition to introducing new knowledge, extend core principles.

Each teaching and learning module will normally be assessed through an associated assessment unit.

Assessment units

AS and A2 biology each comprise six assessment units or modules. In both AS and A2 two of the units are assessed by written examination taken at two specific times of the year, January and June. The third component in each case involves a method of practical assessment. Awarding bodies use either centre-assessed coursework or some form of practical examination.

What are the main differences between AS and A2?

→ AS and A2 courses are designed so that the level of difficulty steadily increases, with the A2 containing the more demanding concepts.
→ In A2 there is a much greater emphasis on the skills of application and analysis than in AS.
→ A2 includes synoptic assessment.

Modules and their shelf-life

Any module can be retaken to improve your grade. You may resit any unit *once only* during the time when the results are held in the Awarding Body's unit bank. When you decide to cash in for an AS/A-level award the Awarding Body will use the best result from each unit that you have attempted. Any AS result can be converted into a full A-level award by taking the A2 exam at any time while the specification remains valid.

Links

While the content of each of the modules at AS is self-contained, during the study of each module opportunities should be taken to enable you to integrate different aspects of the subject. The link feature helps you to achieve this by highlighting connections between different chapters in the revision guide. The specification is designed to allow for the progression from AS to A-level, both in level of difficulty of the topics and in the skill and understanding which are required. Certain topics and concepts which appear as foundation in the AS modules are developed and extended in A2 modules. An example of this is given below:

AS topics	**A2 topics**
molecules . . .	metabolic pathways, e.g. respiration
range of organisms . . .	taxonomy
DNA/gene technology . . .	applications of gene technology

Examiner's secrets

The examination specification (formerly called the syllabus) is an extremely *useful* document. It clearly states the requirements for each topic you study. You should obtain your own copy, from your teacher, the internet, or by application to the relevant examination board (known as Awarding Body). If you are not sure which Awarding Body to contact, your teacher will be able to help you.

Assessment objectives

For AS and A-level the scheme of assessment will test your ability in the following areas:

→ knowledge and understanding
→ application of knowledge and understanding, analysis, synthesis and evaluation
→ experiment and investigation
→ quality of written communication

At A2 there is an additional category: synthesis of knowledge, understanding and skills. This is the synoptic element, where knowledge and skills from different modules are linked together and assessed.

Examiner's secrets

The margin 'Links' feature will certainly help you to make the synoptic links required at A2!

Study skills

This spread gives advice on taking notes during your studies and revision techniques to help you prepare for your exams.

Note taking

Note taking involves condensing speech or writing into an abbreviated form, which nevertheless contains the same *essential* information as the original. This process makes you concentrate on these essentials and in itself is an aid to learning. The second function of note taking is to provide you with an aid to future study and revision. It is much easier to refer back to, or revise from, notes which you have written and organized in your own personal style than from a textbook.

If you are a student attending classes you will probably make most of your notes in the classroom and your teacher will be the main source of information. There are two pitfalls you must try to avoid: do not try to take down every word the teacher says and when making notes from written sources do not copy whole passages. The purpose of note taking is to produce a record of the essential core information.

The following are some points of general advice.

→ Use a loose-leaf file so that you can add to, or rearrange, your notes.
→ Space your notes out on the page so that you can add further information later.
→ Pick out key words or headings by underlining or using colour.
→ Always write down book and page references in case you need to check back at a later date.
→ Consider annotated diagrams as an alternative way of presenting information.

Diagrams

Annotated diagrams are a good way of revising. Starting with a large blank piece of paper can be a good way to focus your mind on a particular topic. You could draw a generalized diagram of a plant in the centre of the paper and then spread more detailed diagrams around it. For example a root hair, a section through the root showing xylem and phloem, an explanation of the cohesion–tension hypothesis of water flow, a section through the stem, a section through the leaf, noting in each case how each organ matches its function. In this way several topics, which may have been studied separately, are brought together. This is an important aspect of revision in the study of biology.

Action point

Make your own notes from several sources. Authors vary in the depth of treatment of topics, therefore it is advisable to read a range of publications.

Examiner's secrets

Teachers recommend using more than one of your senses to aid learning. That's why they often combine 'talk and chalk' or talk with other visual material such as worksheets, TV, etc. Think about using graphic representations of knowledge.

Checkpoint

Watching relevant TV is easy and relaxing *but* without note taking or follow-up it does not do much for your communication skills.

Reading

A-level success depends partly upon the ability to acquire and retain a wide range of knowledge. This is most effectively achieved by reading around the subject; class notes and a textbook alone are inadequate. Try to use more than one textbook as authors vary in their treatment of topics. There are also a number of excellent television programmes on biological topics.

Examiner's secrets

Read around the subject. This increases your vocabulary and grammar.

Action point

The feature 'Check the net' suggests you obtain information using the computer.

Language skills

Frequently, students who know a reasonable amount of biology don't do themselves justice in an exam because of their poor exam technique. Much of this is to do with the problem of language. It can make the difference between passing or failing. You need to be able to read, interpret, memorize, understand and convey information concisely in order to answer exam questions. In many of the older textbooks the reader must interpret the complex language used as well as dealing with unfamiliar biological terms. Modern textbooks put much more emphasis on readability, so most of the unfamiliar words presented to the reader will be of a biological nature.

Examiner's secrets

Your teacher should be used effectively for advice and guidance throughout the course. It is essential that your essays and structured questions are set and marked. Learn to regard all criticism as positive and do not repeat errors.

Understand the basic principles

Many students struggle to remember parts of their subject simply because they do not understand the basic principles involved. Valuable time could be saved by students 'skimming' their notes first to identify such areas in time for them to be dealt with. Facts and relationships are better remembered through a framework of understanding. It is surprising how many students enter exams unaware of what to expect, even to the extent of not having seen a previous year's exam paper beforehand. Every year awarding bodies produce reports on each subject highlighting common misconceptions and providing suggested answers. Your teacher should have a copy of this report. The examiners' comments on candidates' answers are very helpful.

Action point

Use the key words. Making your own notes on the words shown in bold in this guide is a valuable exercise. They can be easily referred to and can be used as 'memory joggers'.

Action point

Organize your time. Prepare an exam revision timetable. Devote equal time to all topics and leave sufficient time to go over the more difficult topics again.

Revision skills

You must revise regularly. Repetition is an important tool in learning and in preparing for exams. It is not a good idea to leave the revision until just before the exam. Revise each topic as soon as it is completed. Success is achieved through consistent work rather than attempting to commit everything to memory at the last minute.

Exam papers

At both AS and A-level there is a considerable emphasis on understanding, interpreting and applying knowledge. In addition to the factual content, you need to appreciate the underlying principles of the subject and understand the concepts and ideas associated with them. This, and the information below, will better prepare you for the types of question you will encounter on exam papers. It will also enable you to understand the principles involved.

Don't forget

The Study Guide concentrates on preparing you for the written exam but your experimental and investigative skills will also be assessed in these papers. Revise your practical work in conjunction with your theory!

Types of question

Students are often presented with information on topics which they have not previously encountered and are asked to apply their general biological knowledge in answering questions. The reasons for this are that examiners are required to set questions of different types which test different ways of thinking. These are as follows.

→ Testing knowledge and understanding: the need to know and recall biological facts, name structures, describe their functions, describe how biological processes are carried out and how they are related to each other.

→ Testing skills and processes: here you are expected to apply your biological knowledge and understanding to new situations, e.g. the handling of numerical data in the form of graphs or tables; the analysis and evaluation of numerical data or written biological information; the interpretation of data; the explanation of experimental results.

→ Testing applications of biology: specifications aim to promote an awareness and appreciation of the significance of biology in 'personal, social, environmental, economic and technological contexts'. This type of question not only tests your understanding of the biological principles involved but also your awareness of some of the arguments surrounding an issue, e.g. the use of genetically modified organisms.

A well-balanced exam paper will test a certain proportion of each of the above skills. You must be prepared not only to learn biological facts but also to use your background understanding to interpret and explain unfamiliar biological data, diagrams or descriptions.

Action point

Do rough calculations of how much time to spend on questions for the papers you are sitting. Use these time allocations when practising answers and develop a confident view of how much you can write in the time.

Exam questions

Each module will contain a number of structured questions with one essay question. There may be a choice of essay questions but no more than one out of two per module. Examiners spend a considerable amount of time preparing, revising and refining exam questions. The contents are agreed upon by a committee after each question has undergone thorough and rigorous analysis to minimize the chance of misinterpretation by the student.

During revision, when confronted with a mass of information, it is tempting to try to work out which topics are 'going to come up' and leave out certain topics. This is not a good idea with modular exams because the questions are designed to cover every topic. You should be asking yourself what you should do with the knowledge that you have.

Structured questions

These are the most common types of questions used in AS and A-level biology exams. They can be short, requiring a one-word response, and/ or can include the opportunity for extended writing. Remember that the number of lined spaces on the exam question paper, together with the mark allocation at the end of each part question, are there to help you. They are indications of the length of answer expected.

Structured questions are in several parts, usually about a common content. You will experience an increase in the degree of difficulty as you work your way through the question. The first part may be simple recall, perhaps defining a term. The most difficult part is at the end, often asking a question relating to 'technological and social applications of biological principles'.

Essay questions

All too often students rush into essay questions, writing everything they know about the topic. If candidates took more time, reading the question carefully, they would have a much better chance of finding out exactly what the examiner requires in the answer. When you start each question, plan it out first and write down your plan. This will not only help you to organize your answer logically but will also give you a checklist to which you can refer while writing your answer. In this way you will be less likely to repeat yourself, wander off the subject or miss out important sections.

Look at this sample question:

> Describe the pathways and mechanisms involved in the movement of carbon dioxide in a plant from the atmosphere to the chloroplast in the leaf.

The words to highlight are *describe, pathways, mechanisms, carbon dioxide, atmosphere, chloroplast*.

The pathway refers to the route taken by the carbon dioxide through stomata, air spaces of the spongy mesophyll layer to the palisade mesophyll cells and into the chloroplasts. The mechanism means 'What makes carbon dioxide take this route?' Carbon dioxide moves in by diffusion because there is more carbon dioxide in the atmosphere than in the chloroplasts during the day since carbon dioxide is being used up in the chloroplasts in the process of photosynthesis.

Key words in examinations

Exam questions use various words that key you in on how to approach the question. A list of the words frequently used in examination questions, together with their approximate meanings, is given below.

Annotate

Give notes of explanation, e.g. each label of a large labelled annotated diagram would include a short description of the function and/or structure as appropriate.

Brief

A short statement of only the main points.

Calculate

Work out, showing all the stages in the derivation of the answer.

Compare

Write about the similarities and differences between two topics, e.g. compare meiosis and mitosis. You need to be organized and without careful planning you will produce a disorganized answer which will be difficult to mark.

Criticize

State the faults or shortcomings of, for example, an experiment.

Define

State the meaning of, for example, a term without actually using the term itself.

Describe

A request for factual detail about a structure or process expressed logically and concisely, e.g. describe the events that occur during the ventilation of the lungs.

Discuss

A critical account of the various viewpoints and arguments in the topic set, drawing attention to their relative importance and significance, e.g. 'Discuss the advantages and disadvantages of biological control'.

Distinguish between

State the differences between, for example, two or more terms, often for the purpose of identification.

Explain

Describe and give reasons for, e.g. explain the concept of autotrophic and heterotrophic nutrition. It is not enough simply to state what these terms mean. You have to explain why some organisms are autotrophs and others are not.

Graphs

When a graph is being interpreted it is essential to relate any changes or trends to its biological context, using data as support where possible.

Illustrate

Include diagrams or drawings as much as possible.

List

A sequence of numbered points one below the other with no explanation.

Name

Write the full name.

Outline

Give only the main points. This means 'don't go into detail'. But if you have learnt the topic thoroughly you may be tempted to waste time by writing too much!

State

A brief concise answer giving no reasons.

Suggest

There may be more than one explanation but as long as yours is reasonable it will gain some marks. This means that the question has no fixed answer and a wide range of reasonable responses is acceptable.

What is meant by

A definition is usually required. The amount of information to be included is dictated by the mark value.

Use the data

Numerical answer required.

Using past questions

You should make use of past questions, set by the Awarding Body whose exams you are taking, as preparation for the exam. Your school or college may provide you with past papers or you can write to the relevant Awarding Body to purchase copies. Awarding bodies also publish the mark schemes used by examiners. These serve as an excellent guide to what is required on any particular topic.

At the end of each spread in this study guide there are exam questions, and the model answer to each question is given at the end of each section. You should attempt each question *without* reference to your notes. After completing the question, mark it yourself. Your wording may not be quite the same but should contain the key words of the answer. If you have not achieved a satisfactory answer, don't be too despondent. Check your notes and rewrite the answer. This way you will improve your understanding and learn from your mistakes. You will probably remember the right answer much better for having gone over it again.

> **Action point**
>
> Practise exam questions. At AS/A-level, knowledge of the facts is not sufficient. You must be able to apply your knowledge to new situations encountered in the exam question.

Exam planning

One of the keys to exam success is to know how marks can be gained or lost. There are two important aspects to this:

→ ensuring that you follow the instructions on the exam paper
→ understanding how papers are marked by examiners

These will help you to maximize your marks when you tackle the exam paper.

The examiner's aim

The examiner's job is to produce a fair exam, consistent in style and comparable with equivalent papers in previous years, which tests the biology contained in the specification.

The examiner is looking for certain points in an answer according to an agreed mark scheme. It is important to write legibly and concisely, although the examiner is primarily looking for biological knowledge and application.

Organize your time

Look at the mark allocation for the question. Try to answer every part of every question. If a part question is worth four marks the first two or three points will be easier to access than the fourth but persevere. Time spent planning an essay is time well spent. If there is a choice it also helps to choose the right question. How often has a candidate started an essay question only to find that he/she does not know as much as they thought about that topic? Planning will highlight what you can remember about a topic and whether you have sufficient information to answer all aspects of the question fully.

Allow a little time to read over the question at the end. You will only do yourself justice if you:

→ take your time
→ read through each question carefully before attempting it
→ answer each question to the best of your ability

Using diagrams

These should always be large, drawn in the correct proportion, fully labelled, have a title and, when appropriate, be referred to in the text of your response. Label lines should be drawn with a ruler so that they do not cross other label lines. They should touch the appropriate structure. Colour and shading should normally be avoided.

There is a difference between a sketch and a diagram. A sketch implies a simple, freehand representation and can be useful to illustrate points relating to experimental situations. Although sketches can be drawn quickly they must be neat; all too frequently they are hurriedly and carelessly drawn. Where the question specifically asks for a diagram, marks will be allocated accordingly in the mark scheme. A diagram must be integrated into, and must make a contribution to, the value of the answer. Detail on the diagram should be limited to the points you wish to make.

→ Don't leave the examiner to work out which part of the diagram is answering the question.
→ Do not rely solely on your labelled diagram.
→ Develop and expand points in your written answer so that the examiner is in no doubt that you realize the relevance of the diagram to your answer.

Using a well-labelled or annotated diagram can be a valuable aid to an exam response. A diagram can often record a considerable amount of information in a short space thus saving several lines of description which in turn saves time. But watch out – don't spend too long on your diagram, particularly if it relates to only part of a question. Try to practise in advance drawing some of the key diagrams you may need.

Using flowcharts

Flowcharts are an excellent aid to revision, used to summarize a topic. However, you cannot answer an exam question solely by using a flow-chart or a metabolic pathway. You should include some descriptive text to explain the purpose of the chart or pathway, making it clear to the examiner what the chart is meant to show.

Synoptic assessment

One of the aims of A2 assessment is to enable candidates to show knowledge and understanding of facts, principles and concepts from different areas of biology and to make connections between them. You must understand what is meant by 'synoptic assessment' if you are to be fully prepared for your A2 examinations.

Synoptic assessment

→ Involves the drawing together of knowledge, understanding and skills learned in different parts of the AS/A2 biology course.

→ May require you to apply knowledge of a number of areas of the course to a particular situation or context.

→ May require you to use knowledge and understanding of principles and contexts in planning experimental work and in the analysis and evaluation of data.

Why do A2 exams contain a synoptic element?

→ Modular exams were introduced for a variety of reasons, one of which was to allow students to study a 'theme' in a manageable block of information.

→ However, in modular courses there is a tendency for candidates to gain success in a module then forget the concepts they have learned as soon as the exam has been taken.

→ Synoptic questions require you to transfer concepts from one module to another and to make connections.

→ Having an overview of a subject is thought to be a better preparation for higher education or employment.

→ Being able to link information gives you a greater understanding and is more satisfying and fulfilling, allowing you to contribute more effectively in problem-solving situations.

Consider the analogy of a computer game program. Writing the various programs to create the characters, scenes and situations is an achievement in itself but seeing how the characters interact in the game as a whole is far more satisfying.

How can I gain synoptic skills?

→ Biology is inherently cyclical and therefore synoptic. During your study of A2 biology you naturally apply many of the ideas and skills learned during AS biology, and even your GCSE course! So in studying A2 biology you will have been using synoptic skills without realizing it.

Differ

asses

→ St
th
qu
ta

→ Es
lin
ex

→
→
→

→ A tems that you already have on
oan

→ Fo
qu

→ Re
lin

→ Ca

→ M
pr

→ As
di
th

Yes, t
asses

→ Ec
→ Po
→ Di
→ Ce
→ Fo
→ Bi

Library @ the Life Centre

Customer ID: **********4199

Title: Biology : A-level study guide
ID: 30119025994775
Due: 11 August 2015

Total items: 1
20/07/2015 13:32

Title: It's up to you, New York : will
Jolly find love in the city where
dreams are made?
ID: 30119027059213
Due: 11 August 2015

Thank you for using
the Library @ the Life Centre.
Please retain this receipt.

For Renewals:
Web: www.sutton-libraries.gov.uk
24 hour phone: 020 8409 7066
Library hours: 020 8770 4700

Follow "Library at The Life Centre"
on Facebook

---

for their synoptic

ased on biological
r this type of
biology and your
le question.
here you make the
s of biology. Some

ency' of the cell?
als.
onment.

s?

will contain synoptic

do so list topics that
A2.

pecimen questions

nich draw together
t and 'check

sily to synoptic

g.

d evolution.

Examiner's secrets

With essay questions you make the
links between different areas of biology
yourself but with structured questions
the context has been chosen for you.

Action point

Check with your teacher, the specification
and look at past papers to determine the
type of synoptic questions you can expect.

Action point

As you review your notes make your own
synoptic links.

Examples of synoptic questions

Examiner's secrets

In part (a) (ii) don't be content with a superficial answer. If a number of lines are provided this suggests that a detailed answer is required.

Examiner's secrets

In addition to the skill of graph interpretation, the question requires a knowledge of taxonomy, mutations, resource management, population growth, pesticide resistance, pest control and biodiversity. It makes links with different concepts but very little depth of knowledge is required.

1 A short structured question

This question uses specific information provided.

The graph below shows the percentage of barn owls found with rat poison in their livers since 1983. The owls are eating resistant rodents (rats and mice) which have eaten the poison, but have not themselves been killed.

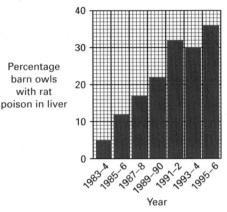

(a) (i) What is the change in percentage of barn owls found with poison in their livers from 1983–4 to 1995–6?

  (ii) How have the rodent populations that are resistant to the poison arisen?

(b) Pesticides are used to kill insects that cause damage to crops. The farmers of Warangal, a district of India, planted large areas of arable land with only cotton for a number of years. The crop is prone to pest attack and disease. *Spodoptera litura* is a caterpillar that attacks cotton. Despite frequent applications of pesticide the caterpillar has continued to destroy the crop and is now eating other vegetable crops as well.

  (i) Name the taxon level represented by *Spodoptera*.

  (ii) What term is applied to the type of crop-growing described?

  (iii) Why does growing a crop on this scale often lead to increased attack by pests?

  (iv) Suggest two ways that the farmers could have altered their farming practices in order to reduce the attack by pests.

Answer to question 1

(a) (i) 36 – 5 = 31%

  (ii) Variations/mutations in the genes of rodents allow them to survive the poison/a mutation has arisen that enables the rodent to produce an enzyme which breaks down the poison; these advantageous genes are passed to the offspring which are then also resistant.

(b) (i) Genus

  (ii) Monoculture

  (iii) Pests have no/limited competition for food; so the reproduction rate is fast/exponential; allowing easy access/ transmission from plant to plant.

  (iv) Plant other crops between the cotton; use biological control/a natural predator; use a number of different pesticides to reduce the chance of the pest becoming resistant to all of them; use integrated pest management. (Any two responses)

2 An essay question

This tests your breadth of knowledge and understanding.

> Discuss the statement that 'Living organisms can acquire and transform energy, using it to synthesize complex molecules'.

Answer to question 2

The following is an outline of the type of answer expected.

Energy from sunlight is captured by the plant using pigments, e.g. chlorophyll, and transformed in the chloroplast to chemical energy, resulting in the production of ATP.

During the light-independent stage of photosynthesis the energy is incorporated into hexose sugar. This may be polymerized into starch. The plant must synthesize all organic materials from the intermediates and products of photosynthesis.

Energy flow through the ecosystem – a brief discussion of trophic levels and food chains, food webs. Energy losses by respiration and excretion. The role of decomposers in the release of energy 'locked' in dead organic material.

Respiration – Krebs cycle and the electron transport chain resulting in the release of energy from glucose in the form of ATP. The use of ATP for active transport, muscle contraction, etc.

Watch out!

All essays, whether synoptic or based on the specialized option within a specific module, usually include marks for correct punctuation, spelling and grammar.

Examiner's secrets

Energy is involved in so many processes that it is very likely to be included in a synoptic question.

Plan the essay first. In this type of essay it is the breadth of knowledge rather than the depth that will gain marks. Don't be tempted to write all you know about respiration and not leave yourself sufficient time to cover the other areas required in the answer.

3 A comprehension question

In this type of question a passage from a scientific journal is used as a basis to test your knowledge of the topic by asking you to explain phrases within the text. At the same time it is a synoptic question in that it gives you the opportunity to show your understanding and make connections between the ecological concepts of populations and competition and the concepts of evolution, selection and speciation. These are linked together by the effect of climatic change on the numbers of plant and animal species and how a change in the balance of populations can affect a species over time.

> Read the following information, which is based on an account in Steve Jones's book on evolution – *Almost Like a Whale* (Anchor Press, 2000).
>
> On one of the Galapagos Islands two types of Darwin's finches are found. One is a large, principally ground-dwelling bird, feeding mainly on hard nuts. The other is smaller and has adapted to feeding on soft fruits and berries. The fossil record shows that both of these two types have been in existence for thousands of years.
>
> Every few years the winds of the Pacific Ocean change direction, creating a current of warm tropical waters, called El Niño, which moves eastwards towards America. In the Galapagos Islands El Niño causes a huge increase in rainfall and a corresponding growth in lush and abundant vegetation. Many fertile hybrid finches are hatched which are intermediate between the large and small forms.
>
> As El Niño ebbs away, drought follows the torrential rain and most of the finches die.

(a) Describe the main adaptation that the larger finch would show to enable it to feed successfully.

(b) (i) The two types of finch are examples of different

Genera   Species   Varieties   (Circle your choice)

(ii) Explain your answer.

(c) (i) What effect would you expect the arrival of El Niño to have on the finch population?

(ii) Explain your answer.

(d) Suggest why 'many hybrids are hatched'.

(e) (i) Under what conditions is competition most severe

normal   El Niño   (Circle your answer)

(ii) Explain your choice.

(f) (i) What does the fossil record tell us about what happens when El Niño 'ebbs away'?

(ii) Explain why this happens.

(iii) Name the evolutionary process this illustrates.

(g) Suggest what might happen to the finch population **over a long period of time**.

(i) I   If El Niño became permanent.

II Explain your answer.

(ii) I   If El Niño disappeared permanently.

II Explain your answer.

Answer to question 3

(a) Strong/more powerful/large (not sharper).

(b) (i) Circle 'varieties'.

    (ii) They can still interbreed; to produce fertile offspring.

(c) (i) The population would increase.

    (ii) There is a large increase in vegetation and so there is more food; the consequence of this increased food supply is that there will be increased breeding success.

(d) Since there is an increase in numbers the two isolated populations are more likely to come in contact with each other.

(e) (i) Circle 'normal'.

    (ii) Under normal drought conditions there is less food available so there is an increase in competition for the reduced food supply.

(f) (i) The intermediate forms do not survive/all the hybrids disappear.

    (ii) The specialized feeders survive because they are best adapted to compete for the limited food supply caused by the drought; the intermediate forms starve. (2 out of the 3 points needed)

    (iii) Natural selection/survival of the fittest.

(g) (i) I   The two populations might merge into one large intermediate group.

        II   With abundant food the specialized feeders might be at a disadvantage.

    (ii) I   Two species would merge, incapable of hybridization.

        II   With no periodic abundance, isolation of two forms becomes permanent.

5 Understanding and interpretation of diagrams

The diagram shows structures that are scattered on the **lower** surface of the leaf and allow the exchange of gases to take place.

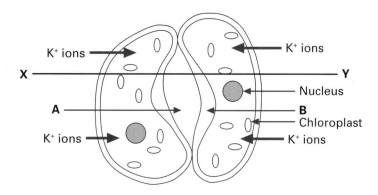

(a) (i) Name the features labelled **A** and **B**.
   (ii) Draw a simple diagram of a transverse section along **X–Y** to show the **relationship** between this structure and the two cell types that lie closest to it.
   Label **each** of the **cell types** shown in your diagram.

(b) One theory suggests that the space labelled A becomes larger as a result of the influx of K^+ ions into the cells shown in the diagram.
   (i) Name the **process** that results in the movement of K^+ ions.
   (ii) Describe how this influx of K^+ ions results in an increase in the size of **A**.
   (iii) **Circle** the line that describes the stage that provides a source of energy for this process.

   Calvin cycle   Cyclic photophosphorylation
   Light-independent reaction   Photolysis

Answer to question 5

(a) (i)   A Stoma/stomatal pore or aperture. B Cell wall.
   (ii) Diagram showing TS lower surface of leaf with labelled cells; two guard cells with epidermal cell on each side and two or three spongy mesophyll cells.

(b) (i)   Active transport.
   (ii) (The influx of K^+ ions results in) The water potential of the guard cells becoming more negative; water enters; by osmosis; the guard cell becomes turgid/increases in volume.
   (iii) Circle 'cyclic photophosphorylation'.

Glossary

Action point

This is a selection of biological words you will encounter and is not a complete list of all the words you are required to know. As you work through each chapter, list the key biological words you come across and write out their definitions.

absorption spectrum
The range of a pigment's ability to absorb various wavelengths of light.

acetyl CoA
The entry compound for the Krebs cycle in cellular respiration.

action potential
A rapid change in the membrane potential of a nerve cell, caused by stimulus-triggered, selective opening and closing of sodium and potassium ion channels.

active site
The specific portion of an enzyme that attaches to the substrate by means of weak chemical bonds.

active transport
Process by which substances are moved across a cell membrane against a concentration gradient, with the help of energy input and specific transport proteins.

adaptation
A feature that increases the chance of survival of an organism in its environment.

adaptive radiation
The emergence of numerous species from a common ancestor introduced into an environment.

allele
An alternative form of a gene.

antibiotic
A chemical that kills bacteria or inhibits their growth.

antibody
An antigen-binding immunoglobin, produced by B-lymphocytes, that attacks non-self proteins in the body.

anticodon
A specialized base triplet on one end of a tRNA molecule that recognizes a particular complementary codon on a mRNA molecule.

antigen
A non-self protein that is recognized by the immune system and elicits an immune response.

artificial selection
The selective breeding of domesticated plants and animals to encourage the occurrence of desirable traits.

Examiner's secrets

Exam questions often ask you to define a term. This is the examiner's way of focusing your mind on a topic. It's also an easy way to pick up marks!

Examiner's secrets

When students first study AS/A2 biology many are surprised that there are so many terms to learn. Expect to define these terms to gain marks in both structured and essay questions in exams.

asexual reproduction
A type of reproduction involving only one parent that produces genetically identical offspring.

autosome
A chromosome that is not directly involved in determining sex, as opposed to the sex chromosomes.

autotroph
An organism that obtains organic food molecules without eating other organisms. Autotrophs use energy from the sun or from the oxidation of inorganic substances to make organic molecules from inorganic ones.

back-cross
A genetic cross that reveals whether an individual with a 'dominant' phenotype is heterozygous or homozygous dominant.

basal metabolic rate
The minimum number of kilocalories a resting animal requires to fuel itself for a given time.

biodiversity
An expression of the number of different species living in a given ecosystem.

biomass
The dry weight of organic matter comprising a group of organisms in a particular habitat.

biological control
The control of a pest by using a control agent, a predator that feeds on the pest (its prey).

biotechnology
The industrial use of living organisms or their components to improve human health and food production.

carbon cycle
The movement of carbon between living organisms and the non-living environment, and involving the processes of photosynthesis, respiration and decomposition.

carcinogen
A chemical agent that causes cancer.

carnivore
An animal that eats other animals.

carotenoids
Accessory pigments in the chloroplasts that absorb wavelengths of light that chlorophyll cannot.

carrier
An individual who is heterozygous at a given genetic locus, with one normal allele and one potentially harmful recessive allele. The heterozygote is phenotypically normal for the character determined by the gene but can pass on the harmful allele to the offspring.

carrying capacity
The maximum population size that can be supported by the available resources.

central nervous system
Part of the nervous system (brain and spinal cord) that connects sensory and motor neurones.

chemiosmosis
The production of ATP using the energy of hydrogen ion gradients across membranes to phosphorylate ADP.

chlorophyll
The green pigment located within the chloroplasts of plants.

chromosome
A thread-like, gene-carrying structure found in the nucleus.

clone
A population of genetically identical individuals or cells.

codon
The basic unit of the genetic code; a three-nucleotide sequence of DNA or mRNA that specifies a particular amino acid or termination signal.

colony (of bacteria)
A cluster of bacterial cells (clone) which arose from a single bacterium.

community
All the organisms that inhabit a particular area; populations of different species that interact.

compensation point
The point at which the rates of photosynthesis and respiration are equal (balanced) in a plant.

competition
The struggle between organisms to obtain resources. Competition can be interspecific (between species) or intraspecific (within species).

conservation
The management of habitats to maintain or restore species diversity and ecosystem function.

counter-current exchange
The opposite flow of adjacent fluids that maximizes transfer rates; blood in the gills of a fish flows in the opposite direction to that in which water passes over the gills, maximizing oxygen uptake and carbon dioxide loss.

crossing over
The reciprocal exchange of genetic material between non-sister chromatids during synapsis of meiosis I.

cytochrome
An iron-containing protein, a component of electron transport chains in chloroplasts and mitochondria.

cytoplasm
The entire contents of the cell, excluding the nucleus, and bounded by the plasma membrane.

decomposer
Organisms, fungi and bacteria that break down dead organic matter to obtain nutrients.

deforestation
The removal of trees from a forest, for timber and/or to obtain land for cultivation.

depolarization
An electrical state in an excitable cell, whereby the inside of the cell is made less negative relative to the outside than at the resting membrane potential.

detritus
Dead organic matter.

diffusion
The passive movement of a substance down a concentration gradient from a region of high concentration to a region of low concentration.

diploid (cell)
A cell containing two sets of chromosomes (2n), one set inherited from each parent.

dominant allele
In a heterozygote, the allele that is fully expressed in the phenotype.

double fertilization
A mechanism of fertilization in flowering plants (Angiosperms) in which two male nuclei unite with two nuclei in the embryo sac to form the zygote and the endosperm.

ecological efficiency
The ratio of net productivity at one trophic level to net productivity at the next lower level.

ecosystem
A level of ecological study that includes all the organisms in a given area as well as the abiotic factors with which they interact.

electron transport chain
A sequence of electron-carrier molecules that shuttle electrons during the redox reactions that release energy used to make ATP.

endangered species
A species that is in danger of extinction throughout all or a significant portion of its range.

enzymes
A class of proteins acting as catalysts, chemical agents that change the rates of chemical reactions without being used up by the reaction.

eukaryotic cell
A cell with a membrane-bound nucleus and membrane-bound organelles.

eutrophication
The artificial 'enrichment' of aquatic habitats (streams, rivers, ponds, lakes) by excess nutrients (e.g. due to run off of fertilizers from farms), resulting in a fall in the oxygen level of the water.

evolution
All the changes that have transformed life on Earth from its earliest beginnings to the diversity that exists today.

extinction
The process by which a species ceases to exist on Earth, e.g. due to a failure to adapt successfully to a changing environment.

facultative anaerobe
An organism that makes ATP by aerobic respiration if oxygen is present but can switch to fermentation under anaerobic conditions.

feedback inhibition
A method of metabolic control in which the end product of a metabolic pathway acts as an inhibitor of an enzyme within that pathway.

fermentation
A process that makes a limited amount of ATP from glucose without an electron transport chain and produces alcohol or lactic acid as an end product.

fertilization
The fusion of male and female gametes to produce a diploid zygote.

food chain
The relationship of organisms in each successive trophic (feeding) level in a community.

food web
Interconnecting food chains in a community.

founder effect
A cause of genetic drift attributable to colonization by a limited number of individuals from a parent population.

gamete
A sex cell containing half the number of chromosomes (haploid) as body cells.

gene
A discrete unit of hereditary information consisting of a specific nucleotide sequence in DNA.

gene pool
The total of all the genes in a population at any one time.

gene probe
A short sequence of DNA used as a 'genetic marker'.

genetic drift
Changes in a gene pool of a small population due to chance.

genetic engineering
The modification of an organism's DNA, usually by the insertion of additional DNA. The changed organism will then have desirable traits (e.g. resistance to disease) which is inherited by its offspring.

genetically modified organism
An organism (e.g. a crop plant) which has had its DNA changed by genetic engineering.

genome
The complete complement of an organism's genes.

genotype
The genetic make-up of an organism.

global warming
The increase in the Earth's temperature due to the build-up of heat trapping 'greenhouse gases' (e.g. carbon dioxide) in the atmosphere.

gross primary productivity
The total primary productivity of an ecosystem.

herbivore
A heterotrophic animal that eats plants.

heterozygous
Having one dominant and one recessive allele for a given characteristic.

homeostasis
The maintenance by organisms of a constant internal environment.

homologous chromosomes
Chromosome pairs of the same length that possess genes for the same characters at corresponding loci. One homologous chromosome is inherited from the father, the other from the mother.

homozygous
Having two identical alleles for a given characteristic.

Human Genome Project
An international collaborative effort to map and sequence the DNA of the entire human genome.

immunity
The ability of an organism to resist disease.

incomplete dominance
A type of inheritance in which the F1 hybrids have an appearance that is intermediate between the phenotypes of the parents.

induced fit
The change in shape of the active site of an enzyme so that it binds more snugly to the substrate, induced by the entry of the substrate.

insulin
A hormone that causes the body to convert excess glucose into glycogen for storage in the liver.

interstitial fluid
Also known as tissue fluid, this forms the internal environment of vertebrates, consisting of the fluid filling the spaces between cells.

***in vitro* fertilization**
Fertilization of the ova in laboratory containers followed by artificial implantation of the early embryo into the mother's uterus.

isotonic solutions
Solutions of equal solute concentration.

Krebs cycle
A chemical cycle that completes the metabolic breakdown of glucose molecules to carbon dioxide with the release of energy.

law of limiting factors
The rate of a process involving several factors will be limited by that factor which is in shortest supply.

lignin
A complex aromatic compound which impregnates the cellulose matrix of plant cell walls, making the wall strong and rigid and impervious to gases, water and solutes.

linked genes
Genes that are located on the same chromosomes.

locus
A particular place along the length of a certain chromosome where a given gene is located.

lymph
A colourless fluid, derived from tissue fluid, found in the lymphatic system of vertebrates.

lymphocyte
A white blood cell. The lymphocytes that complete their development in the bone marrow are called B-cells, and those that mature in the thymus gland are called T-cells.

lytic cycle
A type of viral replication cycle resulting in the release of new phages by death or lysis of the host cell.

meiosis
A two-stage type of cell division in sexually reproducing organisms that results in gametes with half the chromosome number of the original cell.

memory cell
A clone of long-lived lymphocytes, formed during the primary immune response, that remain in a lymph node until activated by exposure to the same antigen that triggered its formation.

metabolism
All the organism's chemical processes, consisting of anabolic and catabolic pathways.

micropropagation
A technique by which individual plants are grown from a cluster of rapidly dividing cells on a nutrient medium, i.e. in tissue culture.

mitosis
A type of cell division that conserves the chromosome number (diploid) by equally allocating replicated chromosomes to each of the daughter nuclei.

monoculture
The cultivation over a large area of a single crop plant.

mutagen
A chemical or physical agent that interacts with DNA and causes a mutation.

mutation
A permanent change in the DNA or the chromosome number of the cell.

natural selection
The differential success in the reproduction of different phenotypes resulting from the interaction of organisms with their environment. It is a process that encourages the transmission of favourable alleles and hinders the transmission of unfavourable ones so contributing to evolution.

negative feedback
A primary mechanism of homeostasis. When there is a change in a monitored physiological variable a response is triggered to counteract the initial fluctuation.

net primary productivity
The gross primary productivity minus the energy used in respiration. It represents the energy in an ecosystem available to consumers.

neurotransmitter
A chemical messenger released from the synaptic terminal of a neuron at a chemical synapse that diffuses across the synaptic cleft and binds to and stimulates the presynaptic cell.

niche
An organism's role in an ecosystem.

nucleus
The chromosome-containing organelle of a eukaryotic cell.

omnivore
A heterotrophic animal that consumes both meat and plant material.

osmoregulation
The control of the water balance in organisms.

osmosis
The diffusion of water molecules across a selectively permeable membrane from a dilute solution to a concentrated solution; water diffuses from a region of high water potential to a region of lower water potential.

oxidative phosphorylation
The production of ATP using energy derived from the redox reactions of an electron transport chain.

oxygen debt
The amount of additional energy required to break down lactic acid, which builds up during anaerobic respiration.

parasite
An organism that obtains nutrients from another living organism, known as the host.

pathogen
A disease-causing microbe.

pesticide
A manufactured chemical used to kill pests in managed environments.

phage
A virus that infects bacteria.

phenotype
The expression of an organism's genotype; the physical and physiological characteristics of an organism.

photophosphorylation
The process of generating ATP from ADP and phosphate by means of a proton-motive force generated by the thylakoid membrane of the chloroplast during the light reactions of photosynthesis.

photosynthesis
The conversion of light energy to chemical energy stored in glucose; the process in green plants by which carbon dioxide and water combine, using light energy, to form glucose and water.

plankton
Microscopic organisms that occur near the surface of oceans, ponds and lakes.

plasmid
A small circular piece of DNA found in bacterial cells.

plasmolysis
A phenomenon occurring in plant cells when placed in a hypertonic solution, resulting in the withdrawal of the cytoplasm and the plasma membrane from the cell wall.

pollination
The transfer of pollen on to the stigma of a carpel by wind or animal carriers.

polymerase chain reaction
A technique for increasing the quantity of DNA *in vitro*.

population
A group of individuals of one species occupying a given unit of space at a given time.

predator
Animals specialized in feeding on another species (prey).

primary productivity
The rate at which light energy is converted to chemical energy of organic compounds by autotrophs in an ecosystem.

protein synthesis
The assembly of proteins from amino acid 'building blocks' in the cytoplasm of cells, under the direction of DNA in the nuclues.

proton-motive force
The potential energy stored in the form of an electrochemical gradient, generated by the pumping of hydrogen ions across biological membranes during chemiosmosis.

proton pump
The active transport mechanism in cell membranes that uses ATP to force hydrogen ions out of a cell and, in the process, generates a membrane potential.

pyramid of biomass
A diagram used to represent the decrease in biomass of organisms at each trophic level in a food chain.

pyramid of energy
A diagram which represents the total energy requirement of each successive trophic level in a food chain.

pyramid of numbers
A diagram used to represent the relative numbers of individuals at each trophic level in a food chain.

recessive allele
The form of a gene which, when inherited, is only expressed in an organism's genotype if no dominant allele is present.

recombinant DNA
An organism's DNA to which a section of DNA from another species has been added in genetic engineering.

refractory period
The short time immediately after an action potential in which the neuron cannot respond to another stimulus, due to an increase in potassium permeability.

resting potential
The membrane potential characteristic of a non-conducting, excitable cell, with the inside of the cell more negative than the outside.

rubisco
Ribulose carboxylase, the enzyme that catalyses the first step of the Calvin cycle.

selective permeability
A property of biological membranes which allows some substances to cross.

sex chromosomes
The pair of chromosomes responsible for determining the sex of an individual.

sex-linked gene
A gene located on a sex chromosome.

sexual reproduction
Reproduction involving two parents, each of which provides a gamete which fuse during fertilization, resulting in offspring that have unique combinations of genes.

sister chromatids
Replicated forms of a chromosome joined together by the centromere and eventually separated during cell division.

sodium–potassium pump
A special transport protein in the plasma membrane of animal cells that transports sodium out of and potassium into the cell against their concentration gradients.

speciation
The origin of a new species in evolution.

species
A group of populations of organisms capable of interbreeding with one another.

species diversity
The number and relative abundance of species in a community.

substrate level phosphorylation
The formation of ATP by directly transferring a phosphate group to ADP from an intermediate substrate in catabolism.

succession
A progressive sequence of changes in the flora and fauna of a region, from pioneer to climax communities.

taxonomy
The branch of biology concerned with naming and classifying the various forms of life.

testcross
The breeding of an organism of unknown genotype with a homozygous recessive individual to determine the unknown genotype.

translocation
The transport of sugars and other substances through phloem cells in plants.

transpiration
The evaporation of water from leaves, resulting in the movement of water up through the xylem vessels (the transpiration stream).

trophic level
A feeding level in a food chain or food web.

tropism
A growth response that results in the curvature of whole plant organs toward or away from stimuli that come from one direction.

vaccine
A harmless variant or derivative of a pathogen that stimulates a host's immune system to produce a defence response against the pathogen.

variation
The differences in characteristics of members of the same species.

water potential
The physical property predicting the direction in which water will flow; the tendency for water to leave a system.

zygote
The diploid product of the fusion of haploid gametes in sexual reproduction.

Further reading

General text books

Biology (1992), Rowland, University of Bath Science 16–19, Nelson Thornes (ISBN 0 17 438425 4)

Biology: Principles and Processes (1993), Roberts, Reiss & Monger, Nelson Thornes (ISBN 0 17 448176 4)

A-level Biology (reprinted 1996), Phillips & Chilton, Oxford Univ. Press (ISBN 0 199 14584 9)

Biological Science (3rd edn 1997), Green, Stout & Taylor, Cambridge Univ. Press (ISBN 0 521 56178 7)

Advanced Biology (2000), Kent, Oxford Univ. Press (ISBN 0 19 914195 9)

Advanced Biology (1997), Jones & Jones, Cambridge Univ. Press (ISBN 0 521 48473 1)

Central Concepts in Biology (1995), Jones *et al.*, Cambridge Univ. Press (ISBN 0 521 48501 0)

Advanced Biology: Principles and Application (2000), McKean & Clegg, Murray (ISBN 0-7195-5078-5)

Advanced Biology (2000), Roberts, Reiss & Monger, Murray (ISBN 0-17-438752-6)

New Understanding Biology for A Level (4th edn 1999), Glenn Toole, Stanley Thornes (ISBN 0 7487 39637)

Cambridge Advanced Science (2007), Mary Jones et al, Cambridge University Press (ISBN 978-0-521-78719-2)

Advanced Biology for You (2003), Gareth Williams, Nelson Thornes (ISBN 0-7487-5298-6)

Nelson Advanced Science: Biology, Molecules and Cells (2003), John Adds et al (ISBN 0 7487-7484-X)

Nelson Advanced Science: Biology, Exchange and Transport, Energy and Ecosystems (2004), John Adds et al (ISBN 13:9780748774876)

Nelson Advanced Science: Biology, Respiration and Coordination (2004), John Adds et al (ISBN 0-74877489-0)

Nelson Advanced Science: Biology, Genetics, Evolution and Biodiversity (2004), John Adds et al (ISBN 0-74877492-0)

Reference books

Advanced Biology Topics: Biodiversity – The Abundance of Life (1997), Chapman & Roberts, Cambridge Univ. Press (ISBN 0 521 57794 2)

Mammals: Structure and Function (1998), Clegg (Illustrated Advanced Biology Series), Murray (ISBN 0 7195 7551 6)

Signs, Symbols and Systematics, the ASE companion to 5–16 Science, The Association For Science Education (ISBN 0 86357 232 4)

Biological Nomenclature (3rd edn 2000), ed. Cadogan, Institute of Biology (ISBN 0 900 49036 5)

Microbiology and Biotechnology (1995), Cadogan & Hanks, Nelson (ISBN 0 174 48227 2)

Microorganisms and Biotechnology, Taylor, University of Bath Science 16–19, Nelson Thornes (ISBN 0 333 48320 0)

Microorganisms, Biotechnology and Disease (1991), Lowrie & Wells, Cambridge Univ. Press (ISBN 0 521 38746 9)

Microbiology and Biotechnology (1994), University of Cambridge Local Examinations Syndicate, Cambridge Univ. Press (ISBN 0 521 42204 3)

Applied Genetics (1991), Hayward, University of Bath Science 16–19, Nelson Thornes (ISBN 0 333 46659 4)

Applied Ecology (1992), Hayward, University of Bath Science 16–19, Nelson Thornes (ISBN 0 174 48187 X)

Genetics and Evolution (1999), Clegg (Illustrated Advanced Biology Series), Murray (ISBN 0 7195 7552 4)

Cell Biology and Genetics (1996), Adds *et al.*, Nelson (ISBN 0 174 48266 3)

Revision Express AS Fast-Track Biology (2001), Rowlands, Pearson (ISBN 0 582 43233 2)

Biology Modular Workbook Series (2006), Health and Disease, Greenwood et al, Biozone (ISBN 1-877329-74-6)

This is not a prescriptive list and any further reading should be carried out after consultation with your teacher.

Index